P9-BHT-003

Implementing the ISO 9000 Series

QUALITY AND RELIABILITY

A Series Edited by

1. Designing for Minimal Maintenance Expense: The Practical Application of Reliability and Maintainability, *Marvin A. Moss*
2. Quality Control for Profit: Second Edition, Revised and Expanded, *Ronald H. Lester, Norbert L. Enrick, and Harry E. Mottley, Jr.*
3. QCPAC: Statistical Quality Control on the IBM PC, *Steven M. Zimmerman and Leo M. Conrad*
4. Quality by Experimental Design, *Thomas B. Barker*
5. Applications of Quality Control in the Service Industry, *A. C. Rosander*
6. Integrated Product Testing and Evaluating: A Systems Approach to Improve Reliability and Quality, Revised Edition, *Harold L. Gilmore and Herbert C. Schwartz*
7. Quality Management Handbook, *edited by Loren Walsh, Ralph Wurster, and Raymond J. Kimber*

8. Statistical Process Control: A Guide for Implementation, *Roger W. Berger and Thomas Hart*
9. Quality Circles: Selected Readings, *edited by Roger W. Berger and David L. Shores*
10. Quality and Productivity for Bankers and Financial Managers, *William J. Latzko*
11. Poor-Quality Cost, *H. James Harrington*
12. Human Resources Management, *edited by Jill P. Kern, John J. Riley, and Louis N. Jones*
13. The Good and the Bad News About Quality, *Edward M. Schrock and Henry L. Lefevre*
14. Engineering Design for Producibility and Reliability, *John W. Priest*
15. Statistical Process Control in Automated Manufacturing, *J. Bert Keats and Norma Faris Hubele*
16. Automated Inspection and Quality Assurance, *Stanley L. Robinson and Richard K. Miller*
17. Defect Prevention: Use of Simple Statistical Tools, *Victor E. Kane*
18. Defect Prevention: Use of Simple Statistical Tools, Solutions Manual, *Victor E. Kane*
19. Purchasing and Quality, *Max McRobb*
20. Specification Writing and Management, *Max McRobb*
21. Quality Function Deployment: A Practitioner's Approach, *James L. Bossert*
22. The Quality Promise, *Lester Jay Wollschlaeger*
23. Statistical Process Control in Manufacturing, *edited by J. Bert Keats and Douglas C. Montgomery*
24. Total Manufacturing Assurance, *Douglas C. Brauer and John Cesarone*
25. Deming's 14 Points Applied to Services, *A. C. Rosander*
26. Evaluation and Control of Measurements, *John Mandel*
27. Achieving Excellence in Business: A Practical Guide to the Total Quality Transformation Process, *Kenneth E. Ebel*
28. Statistical Methods for the Process Industries, *William H. McNeese and Robert A. Klein*
29. Quality Engineering Handbook, *edited by Thomas Pyzdek and Roger W. Berger*
30. Managing for World-Class Quality: A Primer for Executives and Managers, *Edwin S. Shecter*
31. A Leader's Journey to Quality, *Dana M. Cound*
32. ISO 9000: Preparing for Registration, *James L. Lamprecht*
33. Statistical Problem Solving, *Wendell E. Carr*
34. Quality Control for Profit: Gaining the Competitive Edge. Third Edition, Revised and Expanded, *Ronald H. Lester, Norbert L. Enrick, and Harry E. Mottley, Jr.*
35. Probability and Its Applications for Engineers, *David H. Evans*
36. An Introduction to Quality Control for the Apparel Industry, *Pradip V. Mehta*

37. Total Engineering Quality Management, *Ronald J. Cottman*
38. Ensuring Software Reliability, *Ann Marie Neufelder*
39. Guidelines for Laboratory Quality Auditing, *Donald C. Singer and Ronald P. Upton*
40. Implementing the ISO 9000 Series, *James L. Lamprecht*

ADDITIONAL VOLUMES IN PREPARATION

Reliability Improvement with Design of Experiments, *Lloyd W. Condra*

Quality by Experimental Design, Second Edition, Revised and Expanded, *Thomas B. Barker*

Implementing the ISO 9000 Series

James L. Lamprecht

International Consultant
Issaquah, Washington

Marcel Dekker, Inc. New York • Basel • Hong Kong

Library of Congress Cataloging-in-Publication Data

Lamprecht, James L.
Implementing the ISO 9000 series / James L. Lamprecht.
p. cm. -- (Quality and reliability ; 40)
Includes bibliographical references and index.
ISBN 0-8247-9134-7 (acid-free paper)
1. Quality control--Standards. 2. Quality assurance. I. Title.
II. Series.
TS156.L32 1993
658.5'62--dc20 92-47141
CIP

This book is printed on acid-free paper.

MARCEL DEKKER, INC.
270 Madison Avenue, New York, New York 10016

Current printing (last digit):
10 9 8 7 6 5 4 3

PRINTED IN THE UNITED STATES OF AMERICA

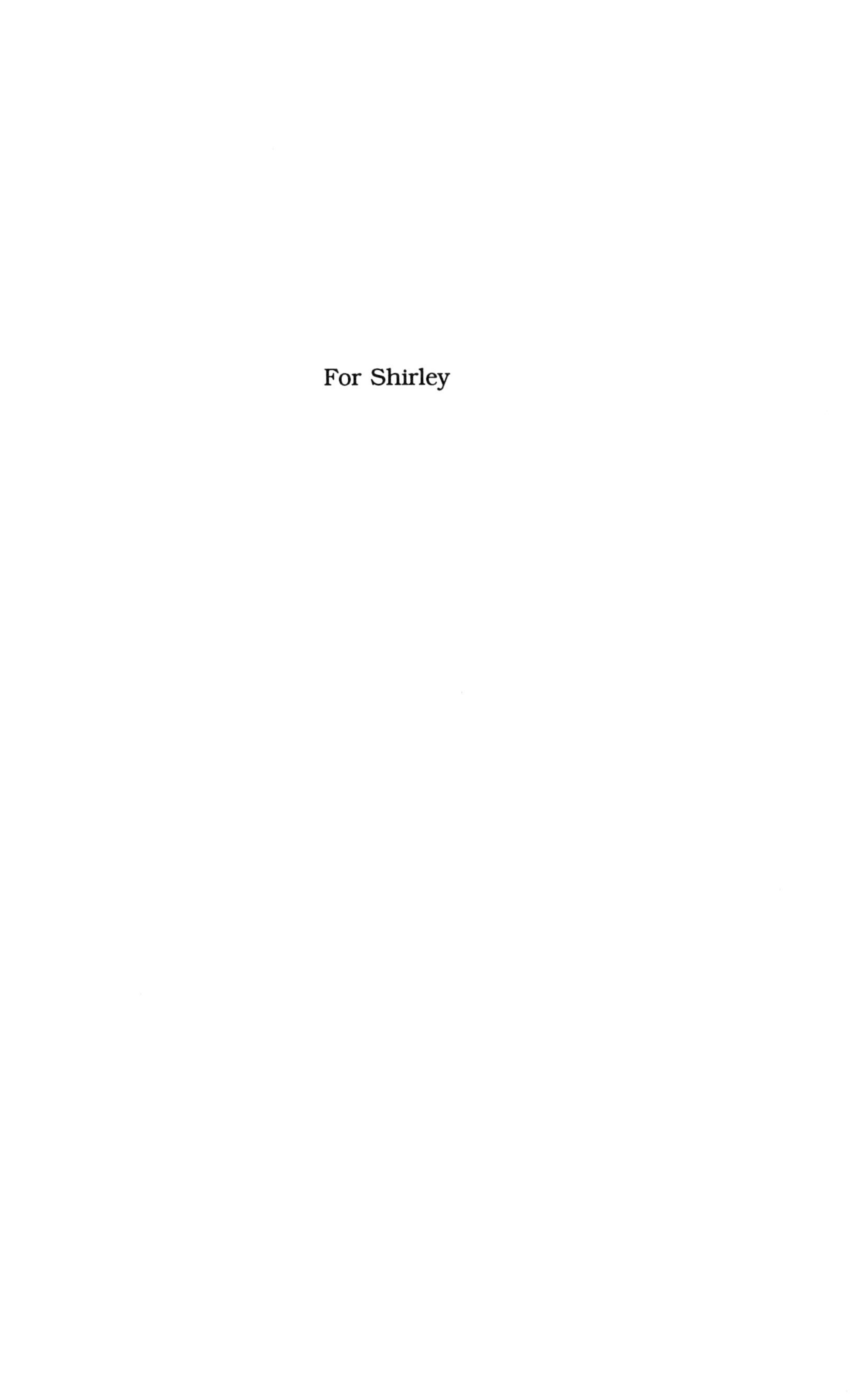

For Shirley

About the Series

The genesis of modern methods of quality and reliability will be found in a simple memo dated May 16, 1924, in which Walter A. Shewhart proposed the control chart for the analysis of inspection data. This led to a broadening of the concept of inspection from emphasis on detection and correction of defective material to control of quality through analysis and prevention of quality problems. Subsequent concern for product performance in the hands of the user stimulated development of the systems and techniques of reliability. Emphasis on the consumer as the ultimate judge of quality serves as the catalyst to bring about the integration of the methodology of quality with that of reliability. Thus, the innovations that came out of the control chart spawned a philosophy of control of quality and reliability that has come to include not only the methodology of the statistical sciences and engineering, but also the use of appropriate management methods together with various motivational procedures in a concerted effort dedicated to quality improvement.

This series is intended to provide a vehicle to foster interaction of the elements of the modern approach to quality, including statistical applications, quality and reliability engineering, management, and motivational aspects. It is a forum in which the subject matter of these various areas can be brought together to allow for effective integration of appropriate techniques. This will promote the true benefit of each, which can be achieved only through their interaction. In this sense, the whole of quality and reliability is greater than the sum of its parts, as each element augments the others.

The contributors to this series have been encouraged to discuss fundamental concepts as well as methodology, technology, and procedures at the leading edge of the discipline. Thus, new concepts are placed in proper perspective in these evolving disciplines. The series is intended for those in manufacturing, engineering, and marketing and management, as well as the consuming public, all of whom have an interest and stake in the improvement and maintenance of quality and reliability in the products and services that are the lifeblood of the economic system.

The modern approach to quality and reliability concerns excellence: excellence when the product is designed, excellence when the product is made, excellence as the product is used, and excellence throughout its lifetime. But excellence does not result without effort, and products and services of superior quality and reliability require an appropriate combination of statistical, engineering, management, and motivational effort. This effort can be directed for maximum benefit only in light of timely knowledge of approaches and methods that have been developed and are available in these areas of expertise. Within the volumes of this series, the reader will find the means to create, control, correct, and improve quality and reliability in ways that are cost effective, that enhance productivity, and create a motivational atmosphere that is harmonious and constructive. It is dedicated to that end and to the readers whose study of quality and reliability will lead to greater understanding of their products, their processes, their workplaces, and themselves.

Edward G. Schilling

Preface

When *ISO 9000: Preparing for Registration* was first published, I naively assumed that most questions had been answered. I soon realized how wrong I was when people kept asking me for more information, interpretations and, above all, examples on how to organize a quality assurance system that would conform to the ISO 9000/ANSI ASQC Q-9000 series.[1] As I found myself answering an increasing number of phone calls and typing a seemingly unending number of three-to-five page letters and memos, the need for a second book became clearly evident.

The challenge facing me was to attempt to expand on the first book without duplicating its contents. Yet, I could not write a workbook that would constantly refer to my first book without running the risk of irritating first-time readers. I was thus faced with the task of offering new material while at the same time including enough information to make the (work)book self-sufficient. This turned out to be more challenging than I had expected. Indeed, although the original intent was to include examples with detailed commentaries, I found myself simultaneously expanding on related themes and offering narratives on several of the ISO 9001 clauses. The workbook took a different form.

Among the many questions that are asked about the ISO series, I have noticed that for some people implementing an ISO quality assurance system needs to be a bit like reading a recipe. "Could you tell me specifically what I must do to satisfy the auditor?" is the type of question that reminds me of the recipe analogy. In other words, "tell me how many spoonfuls of traceability I really need, how many grams

---

[1] James Lamprecht, *ISO 9000: Preparing for Registration* (New York: Marcel Dekker/ASQC Press, 1992). In the U.S., the series is still known as the Q-90 series; however, to avoid confusion, the series was renumbered in 1992 as Q-9000.

of document control and how long must I bake the whole thing in order to satisfy the third-party audit?" I am still perplexed by these questions for the simple reason that the answer can only be "it depends on the nature of your business." There are of course many individuals/consultants/experts who are always eager to tell you how your exact (or nearly so) question was addressed while they worked for a military contractor. Referring to MIL Q 9858, they will tell you that in this particular instance they had to do so and so. Of course, if you should remind them that you never have supplied to the Department of Defense, they will likely reply, "It does not matter, it's all the same!" Well, *it is not*. The ISO 9000 series *is not Mil Q 9858*; it is derived from it, it is very much influenced by 9858, but it should not be interpreted or implemented as 9858. Unfortunately, the same scenario could have unfolded if the "consultant" in question had experience with the nuclear industry or the automotive industry. It is simply amazing how many people interpret the *Statistical Techniques* paragraph as meaning that "you must implement Statistical Process Control." Such recommendations invariably come from individuals (turned consultants) who have had to implement SPC at their plant (usually because their customers requested it, whether it was necessary or not).

There are many experts on the ISO 9000 series; however, these experts come in many grades. While visiting a prospective client in September of 1992, I was asked to view a video tape which had recently been purchased by someone within the company. As my host was placing the tape within the VCR he exclaimed, "I don't think they know what they are talking about, could you please tell me if I am right?" Sure enough, the tape contained much misinformation. The speaker constantly referred to items listed under the ISO 9004 guidelines as if they were ISO requirements. "You will have to maintain a cost of quality program. . . . you will have to implement SPC." That is simply not true. Although the ISO 9000 series may one day require or (more likely) suggest that "when appropriate or if called for by your customers" you shall implement a cost of quality program and SPC, you do not *have* to implement such a program and will likely not have to for at least another ten years. This is not to say that such programs are worthless, for indeed they may well be valuable and beneficial to your company. However, to *suggest* that you might want to implement a cost of quality program or SPC (or any other technique) is not the same as stating that the ISO 9000 series *requires* you to implement such programs. The first half of the last sentence might be good advice, the second half is simply very misleading and plain wrong.

There is at least one more myth surrounding the ISO 9000 series, and that is that some registrars are the self-proclaimed experts on the series. Such claims are of course misleading. It is true that certain European registrars have been registering companies to the BS 5750 series (U.K.'s equivalent to the ISO 9000 series) ever since the mid 1980s. However, that is not tantamount to being an expert on quality assurance systems. In fact, if we are to use length of time in service as a surrogate for expertise, one must then acknowledge the RvC of the Netherlands, which has accredited registrars since 1982. The persistent claim that the center of expertise regarding the ISO 9000 series originated in a particular country should be challenged, for the claim apparently knows no bound and, if unchecked, may well permeate every aspect of quality. A case in point is the attempt by one European registrar to impose its software certification scheme known as TickIT. Already, in an attempt to counter or react to the TickIT program, the Registrar Accreditation Board (RAB) in Milwaukee announced during the last quarter of 1992 the formation of the Software Quality Systems Registration (SQSR) committee. The committee's task will be to study the need for a registration and accreditation system for the application of ANSI/ASQC Q91 (ISO 9001) to software. Given the abundance of software standards, the task will no doubt be complicated.[1]

---

[1]Leonard L. Tripp lists 26 standards relating to the software industry. These include ISO, BSI, IEEE, MIL, CSA, DOD, DIN, AQAP, ANSI and FIPS standards on configuration management, quality assurance and verification and validation standards. See Mr. Tripp's "Software Engineering Standards Today and Tomorrow," in Software Quality, Vol. 5, no. 1 (1992), pp. 4-5. See also Bellcore's

In view of the persisting myths or misinformation about what the ISO 9000 series is or is not, who the experts are, etc., I thought it would be a good idea to expand on some the themes presented in my first book. Rather than simply narrate the implementation process, I have decided to include some information on the European directives; what they are, and who might be affected by them. Also, instead of focusing on the needs of two or three industries, I have provided the reader with as many examples as possible from a variety of industries including the service industry (see particularly Chapter 3). I also thought it would be interesting to occasionally include historical references that directly relate to today's questions. Indeed, many of the requirements stated within the ISO standards, and the subsequent attempts by industry to address these requirements, have their roots in the early days of industrialization.

The reader should be warned that there is no one way to implement an ISO quality assurance system. Realizing that the following comments might upset or distress some readers, one must recognize that implementing ISO within any company cannot be compared to solving a set of equations where you systematically solve for each unknown and *presto* (if you don't have more unknowns than equations), the solution appears. Implementing ISO requires perseverance, commitment, teamwork, planning and artfulness.

Chapter 1 is an overview of some of the important issues relating to the ISO Standards and the major European directives. Chapter 2 looks at the ISO 9000 series as a system that fits within the total quality model. Chapter 3 interprets some of the ISO clauses from various industry viewpoints including the service industries. Chapter 4 focuses on how to bring about change within a company. Chapter 5 looks at various ways to organize a quality assurance system. Chapter 6 offers some suggestions and examples on how to document the elusive tier two. Chapter 7 is a detailed analysis on how to conduct internal

---

Software Quality Program Generic Requirements (SQPR), TR-TSY-000179, 1, July 1989.

audits; it includes several sample forms. Chapter 8 explores some important themes relating to third-party audits and includes two case studies. Chapter 9 reviews the draft updates of the 1987 edition (to be published in late 1993 or early 1994). Chapter 10 covers several themes but basically attempts to answer some questions relating to the future of the ISO 9000 series. The Appendices (A-H) include much information including: addresses and phone numbers of governmental agencies specializing in acquiring information about the ISO 9000 series and the European directives; a partial list of registrars; a sample quality manual; answers to two case studies; a list of ISO/IEC Guides, and more. In addition, the regional addresses of all Trade Adjustment Assistance Centers, which can provide assistance for ISO consultation for certain companies, are also included in Appendix A.

Although most of the companies I have worked with have fewer than 3,000 employees, the suggestions contained within the following pages should apply to any company, large or small (fewer than 25 employees). Naturally, the complexity of the task increases as more divisions are included, but that usually means that much more time needs to be devoted to planning.

Intended audience: The book addresses itself to anyone in need of implementing an ISO 9000 quality assurance system. This may include:

- Companies (small or large) or individuals who would like to implement an ISO 9000 quality assurance system but who, for various (usually ill-advised) reasons, do not wish to invest the time and thus subcontract the documentation to a third party.

- Companies or individuals who have begun implementing ISO but are bogged down in the documentation quagmire.

- Companies or individuals who think they might need to achieve registration but don't quite know what is involved.

- Individuals who have attended one of the many two-day ISO 9000 seminars offered throughout the country and still don't know what is required.

- Companies or individuals who proudly advertise the fact that they have achieved registration *without* the help of a consultant. They might still learn a thing or two.

- ISO 9000 auditors who are charged with the very difficult task of assessing the quality assurance system of a vast array of companies.

- Students who might be enrolled in one of the few quality management programs slowly mushrooming across the nation.

As was the case in my first book, the focus is on ISO 9001. Nevertheless, ISO 9002 and ISO 9003 applicants (few as they are) should also benefit from the book's contents (see Table 3 in Chapter 1 for cross-referencing between the three standards). (Note: It should be stated that some registrars have recently opted to no longer register companies to the ISO 9003 standard *if value added is performed by the company*. These companies will have to seek ISO 9002 registration.) If you understand the ISO 9001 standard, implementing 9002 or 9003 will be easier, since they are subsets of the 9001 document. Even though the ISO 9000 series is equivalent to the European EN 29000 series and the ANSI/ASQC Q 9000 series, I have chosen to refer to the standards as the "ISO 9000 series"; this is mostly out of habit.

As is the case with almost anything relating to the ISO series, subjectivity is of the essence. Thus, contrary to what is sometimes favored by certain experts, the propositions that follow are not intended to be dogmatic. Consequently, readers are encouraged to adopt and adapt what they consider to be appropriate to their industrial environment and customer requirements. I hope the following pages will help you facilitate a successful implementation.

Acknowledgments

I would like to express my sincere appreciation to the following individuals for offering their valuable and constructive criticism: Donald Glen Mason of Petrolite in Tulsa, Oklahoma, Ronald Muldoon of Brown & Root in Houston, Texas and Wendy Schechter of Marcel Dekker, Inc. Naturally, any inaccuracies, inadvertent as they may be, are the author's responsibility. I also would like to thank my wife Shirley for putting up with me during my brief—or nearly so—periods of intense frustration generally induced by an occasionally uncooperative word processor. Her support throughout this project is greatly appreciated.

James L. Lamprecht

Table of Contents

About the Series v

Preface vii

1 European Community Directives and the International Standard Organization 9000 Series 1

The International Standard Organization (ISO) 9000 Series 2
What is the ISO 9000 Series? 4
Some Clarifications Regarding the ISO 9000 Series 9
The EC's Technical Harmonization Directives Regarding Regulated Products 11
Technical Harmonization Directives—General Guidelines 12
European Directives 15
Certification Options (Modules) and the CE Mark 15
Summary and Conclusions 20

2 The ISO 9000 Series as a System 23

What Do You Mean by Total Quality? 24
The ISO 9000 Series as a System 25
What to Consider When Implementing an ISO 9001, 2 or 3 Quality System 27
The Cybernetics of ISO Implementation 30
How to Bring About Order Consistent with the "Show Me Mapping" 32
Summary and Conclusions 35

3 How Do You Interpret This Paragraph? 36

Introduction 36
Does This Paragraph Apply to My Industry? 36
A Typology of ISO 9000 Applicants 37
Types of Processes 40
What Do You Mean by Contract Review? 41
What Do You Mean by Nonconformities? (Review and Disposition) 44
Design and Document Control (4.4) 46
Document Control (4.5.1 and 4.5.2) 49
Process Control (4.9) 52
How and What to Document, and Who Shall Do the Writing? 54
The Generic Process Flow Diagram 56
Inspection, Measuring, and Test Equipment (4.11) 59

Quality Records (4.16) 66
Internal Quality Audits (4.17) 67
Training (4.18) 67
Conclusions 70

4 How to Implement Change 71

On Change 71
Types of Changes: First- and Second-Order Change 72
Second-Order Change 73
How to Bring About Second-Order Change 77
Bringing About Change 78
On Leadership 79
Conclusions—Putting All Together OR "Where Do We Start?" 81

5 The Quality Assurance System—Which Structure to Adopt? 85

Which Structure to Adopt? 86
Which Strategy? 88
How to Organize Tier Two Documentation 93
Suggestions on How to Bring About Implementation 98
Conclusions 102

6 Sample Documentation 104

Multi-Departmental Procedures 119
Conclusion 120

7 Internal Quality Audits 121

Size of the Internal Audit Team 123
How to Conduct Internal Audits 123
Preparation 123
Familiarity with the Standard 127
Human Relations 128
Reporting 128
Closure 129
Samples 129
Sample 1 131
Sample 2 133
Sample 3 134
Sample 4 136
Sample 5 138
Sample 6 139
Conclusions 142

8 A Look at Third-Party Audits **143**

Do I Have to Answer That Question? 144
How to Formulate Questions 146
Case Study/Exercise 1 (Allow 45-60 Minutes) 147
Case Study/Exercise 2 (Allow 45-50 Minutes) 149
The Audit Process—How Reliable and Repeatable? 152
Sources of Variability 154
Auditor's Psychological Types 154
Auditor's Training/Experience 156
On Variability and the Need for Standardization 157
Most Often Made Mistakes When Audited 163
Conclusions 164

9 A Look at the Draft Updates to the ISO 9000 Series **166**

The ISO 9001 Draft Updates 166
A Look at ISO 9002 and 9003 Updates 171
ISO 9000 and ISO 9004 171
Conclusions 172

10 The Future of ISO 9000 .... **174**
Third-Party Audits: Can They Be Standardized? 175
The Political Economy of ISO 9000 178
The Dangers of Institutionalizing the ISO 9000 Series 180
ISO 9000 and Litigations: Real or Perceived Threat? 182
ISO 9000 and Innovation: Common Sense or Contradiction? 184
How Open Is the TC 176 Committee? 187
Did You Say Product Certification? 189
Will the ISO 9000 Series Make It to the 21st Century? 191
Conclusions 193

Afterword **196**

Bibliography **200**

Appendix A: A Look at Some Directives—Telecommunications, Medical Devices and Food and Beverages **205**

Telecommunications 205
Medical Devices 206
Food and Beverages Directives 207
Table A.1 Trade Adjustment Assistance Centers 210

Appendix B: Acromyms **211**

Appendix C: National and International Registrars (May 1992) **212**

Appendix D: Quality Manual **215**
Sample Pages From a Generic Quality Manual 218
QUALITY MANUAL 223
A Comment Regarding Mission Statements 228

Appendix E: Suggested Answers to Case Studies **248**

Appendix F: ISO/IEC Guides **252**

Appendix G: Summary of the "Working Document on Negotiations with Third Countries Concerning the Mutual Recognition of Conformity Assessment" **254**

Appendix H: Component Types Covered by IECQ **256**

Index **258**

Implementing the ISO 9000 Series

1 European Community Directives and the International Standard Organization 9000 Series

The European Summit held December 9-10, 1991 in the small Dutch town of Maastricht emphasized the resolve of some European leaders to achieve a new economic and monetary union by the end of this century.[1] It was perhaps logical that long after Jean Monnet proposed his concept of a European Community, another Frenchman, Jacques Delors (the current President of the European Community Commission), was able to preside over this most important summit and witness the confirmation of ideas which he had proposed over the past decade. One of Delors' most significant contributions to the "New Europe" is the concept of a Single European Market.[2] Starting on January 1, 1993, the Single European Market will allow exporters to the EC who comply with certain directives concerning regulated products, to ship their products to any EC member state through any single gateway (EC) port and hence move freely through the Community.[3]

Most of the new programs that have taken effect starting on January 1, 1993 find their origin in the European Community (EC) July 1985 Product Liability Directive and the Resolution of May 7, 1985 concerning technical harmonization and standardization. As product directives and technical harmonization resolutions are being written, U.S. exporters and potential exporters to the EC have expressed an

---

[1] It is important to note that European governments are not all in agreement as to the need for a monetary union. The recent "No to Maastricht" vote by Danish voters and the surprising anti-Maastricht sentiment of French voters indicates that European economic and monetary unions are not a *fait accompli*. See, for example, "French jitters at growing 'non' vote" in *The European* of the week ending 13 August 1992. *The European* is a very good weekly for European news. It is available in most major U.S. cities.

[2] Within the next five years, Europe's Single Market will include most European Free Trade Association member states.

[3] See Don Linville, "Exporting to the European Community," *Business America*, February 24, 1992, pp. 18-20. It remains to be seen how the system will function and operate within the next few years.

insatiable appetite for information and clarification as to whether or not these new regulations will directly or indirectly affect them.[1]

There are basically two major issues to consider: those relating to the product liability directives and those relating to the technical harmonization resolutions. The Product Liability Directives (also known as Single Market Directives) basically state that for certain *regulated* products, manufacturers exporting to the EC and, eventually, to the European Free Trade Association (EFTA), *will need to have (at a minimum) a well-documented and implemented quality assurance system.* The quality assurance system will need to be modelled according to the EN 29000 series or the national equivalent, e.g., the ANSI/ASQC Q-9000 series. In addition, for certain EC regulated products, the manufacturer will also have to affix a stamp known as the CE mark which would guarantee that the product conforms to certain technical requirements. Before examining the many complex rules and regulations relating to the exportation of EC regulated products, let us first examine the quality assurance requirements as stated in the EN 29000/ISO 9000 series.

The International Standard Organization (ISO) 9000 Series

The European Community's Product Liability Directive of July 1985 states that manufacturers and exporters of a product or service to EC countries "will be liable, regardless of fault or negligence, if a person is harmed or an object is damaged by a faulty (defective) product."[2] To ensure a minimum compliance, many EC directives either require or suggest that the manufacturer shall/should implement, maintain and register a quality assurance system which conforms to one of the European Norms for quality assurance known as EN 29001, 29002 or

---

[1] For a good analysis and historical analysis of the European Community, see Timothy M. Devinney and William C. Hightower, *European Markets After 1992* (Lexington, MA: Lexington Books, 1991) and Rick Arons, *EuroMarketing* (Chicago: Probus Publishing Company, 1991). Arons focuses on marketing strategies and analysis.

[2] Walter H. Boehling, "Europe 1991: Its Effect on International Standards," *Quality Progress*, July 1991, p. 29.

29003 (these standards are equivalent to ISO 9001, 9002 and 9003, the ANSI/ASQC Q 9000 series, and the Canadian CAN/CSA-Q9000 series)[1]. There is in fact, as shall be explained starting on page 12, a variety of options available to U.S. manufacturers.

Founded in 1946, the International Organization for Standardization (commonly referred to as ISO, rather than IOS), consists of some 90 member countries. U.S. participation in the ISO has been through the American National Standards Institute (ANSI). With the exception of electrical and electronic engineering, which are covered by the International Electrotechnical Commission (IEC), the ISO is responsible for the promotion and development of international standards and related activities including conformity assessments such as testing, inspection, laboratory accreditation, certification and quality assessment.[2] It is this last requirement of quality assessment which is of interest to most American manufacturers.

The ISO 9000 series has been adopted by some forty-three countries, including the United States where the series is known as the ANSI/ ASQC (American Society for Quality Control) Q-9000 Series. In Europe, where the standards are known by various acronyms including the European Norm (EN) 29000 series, the ISO 9000 series has been adopted by the European Committee for Standardization (known by its French acronym of CEN) and the European Committee for Electrotechnical Standardization (CENELEC).[3]

---

[1] The Canadian Standards Association (CSA) standards also include Z299 supplement covering unique Canadian requirements.

[2] The reader is referred to Appendix H for a list of electronic components requiring certification by the IEC. For further information write to the Electronic Components Certification Board, Electronic Insdustries Association, 2001 Pennsylvania Avenue N.W., Washington, D.C. 20006.

[3] Several European countries still maintain their own national nomenclature. In the U.K., for example, the EN 29000 series is know as BS 5750 Parts 1-3.

What Is the ISO 9000 Series?[1]

Developed by the ISO Technical Committee (TC) 176, whose members are mostly Americans, Canadians and Western Europeans (particularly British, French, German and Dutch nationals), the ISO 9000 series, published in 1987 and updated approximately every five years, consists of five documents whose focus is quality assurance systems: ISO 9000, ISO 9001, ISO 9002, ISO 9003 and ISO 9004 (see Tables 1.1, 1.2 and 1.3).[2] Of the five documents, two (ISO 9000 and ISO 9004) are *guidelines* and are intended to be used as interpretive references only.

The three standards of relevance to the majority of U.S. industries (i.e., 9001, 9002 and 9003) represent "three distinct forms of functional or organizational capability suitable for **two-party contractual purposes"** (emphasis added). [3]

The *ISO 9001 (ANSI/ASQC Q9001)* is to be used "when conformance to specified requirements is to be assured by the supplier during several stages which *may* include design/development, production, installation, and servicing."[4]

The *ISO 9002 (ANSI/ASQC Q9002)* is to be used when "conformance to specified requirements is to be assured by the supplier during production and installation."[1]

Finally, the *ISO 9003 (ANSI/ASQC Q9003)* is to be used when conformance "to specified requirements is to be assured by the supplier solely at final inspection and test."[1]

---

[1] It is interesting to note that, according to ISO Secretary-General Larry Eicher, the letters ISO do not stand for anything. ISO is simply the prefix meaning equal (e.g. isobars, isotherms, etc.). See *Quality*, October, 1992 p. 21.

[2] The 1992 updates may be published in late 1993.

[3] ANSI/ASQC Q91-1987, p. 1. To avoid unnecessary confusion, the ANSI/ASQC series was renumbered (first quarter of 1992) as the Q- 9000 series.

[4] ANSI/ASQC Q91 *op. cit.*

Each standard (9001, 9002 and 9003) is subdivided into major clauses, which are in turn subdivided into one or more sub-clauses (see Tables 1.4 and 1.5).

Table 1.1 Existing 9000 Series Standards

- 9000: Quality Management and Quality Assurance Standards — Guidelines for Selection and Use.
- ***9001: Quality Systems—Model for Quality Assurance in Design, Production Installation, and Servicing.***
- ***9002: Quality Systems—Model for Quality Assurance in Production and Installation.***
- ***9003: Quality Systems—Model for Quality Assurance in Final Inspection and Test.***
- 9004: Quality Management and Quality System Elements—Guidelines.

Table 1.2 New and Related ISO Standards*

- 10011-1: Guidelines for Auditing Quality Systems, Part 1—Auditing
- 10011-2: Guidelines for Auditing Quality Systems, Part 2—Qualification Criteria for Auditors
- 10011-3: Guidelines for Auditing Quality Systems, Part 3—Managing Audit Programs
- 10012-1: Quality Assurance Requirements for Measuring Equipment, Part 1—Management of Measuring Equipment
- 9000-3: Guidelines for the Application of ISO 9001 to the Development, Supply and Maintenance of Software
- 9004-2: Quality Management and Quality System Elements, Part 2—Guidelines for Services

* Additonal draft documents are currently under consideration, they are: Guidelines for Quality Improvement (9004-4), Guidelines for Quality Plans (9004-5) and Guidelines for Configuration Management (9004-6).

Table 1.3 ISO 45000 Series

- EN 45001: General Criteria for the Operation of Testing Laboratories
- EN 45002: General Criteria for the Assessment of Testing Laboratories
- EN 45003: General Criteria for Laboratory Accreditation Bodies
- EN 45011: General Criteria for Certification Bodies Operating Product Certification
- EN 45012: General Criteria for Certification Bodies Operating Quality Systems Certification
- EN 45013: General Criteria for Certification Bodies Operating Certification of Personnel
- EN 45014: General Criteria for Supplier's Declaration of Conformity

The nested interrelationship between 9001, 9002 and 9003 is revealed in Table 1.5. Indeed, although the numbering of each clause is not yet standardized across the standards (it will be in the next edition), the clauses' title reveal that the twelve clauses of 9003 are a subset of the eighteen 9002 clauses, which are in turn contained within the twenty clauses of 9001. Presently, the decision as to which standard applies to your company or divisions is up to you, the user. However, as more and more product directives are written, the selection might become more focused. Nonetheless, since there are only three standards to chose from and since ISO 9003 only applies to final inspection and testing organizations, the choice is conveniently narrowed down to either 9002 or 9001. If design is an important part of your business activities you should focus on 9001, the most comprehensive ISO quality assurance model. If, however, you wish to register a chemical plant or an assembly plant where no design activities are performed, you might want to consider 9002. In most cases, companies select 9002 for certain plants within a division and 9001 for one or more facilities. If you should need to register several plants, it is highly recommended that you allow each plant to pursue its own registration. Registering several plants under one certificate can lead to some difficulties. For example, if you

Table 1.4 Cross-Reference List of Quality System Elements
(Adapted from the Annex found in ISO 9000, p. 6)

Title	Corresponding Paragraph (or Subsection) Nos. in		
	9001	9002	9003
Management Responsibility	4.1	4.1a	4.1b
Quality System Principles	4.2	4.2	4.2a
Contract Review	4.3	4.3	-
Design Control	4.4	-	-
Document Control	4.5	4.4	4.3a
Purchasing	4.6	4.5	-
Purchaser Supplier Product	4.7	4.6	-
Product Identification + Traceability	4.8	4.7	4.4a
Control of Production	4.9	4.8	-
Inspection and Testing	4.10	4.9	4.5a
Inspection, Measuring and Test Equip.	4.11	4.10	4.6a
Inspection and Test Status	4.12	4.11	4.7a
Control of Nonconforming Product	4.13	4.12	4.8a
Corrective Action	4.14	4.13	-
Handling, Storage, Packaging + Delvry	4.15	4.14	4.9a
Quality Records	4.16	4.15	4.10a
Internal Audits	4.17	4.16a	-
Training	4.18	4.17a	4.1b
After-Sales Servicing	4.19	-	-
Statistical Techniques	4.20	4.18	4.12a

Key

a **Less stringent than 9001**

b **Less stringent than 9002**

- **Element not present**

Unmarked paragraphs indicate Full Requirement

Table 1.5 ISO 9001/Q91 Sub-Heading Lists

4.1 Management Responsibility

4.1.1 Quality Policy
4.1.2.1 Responsibility and Authority
4.1.2.2 Verification Resources and Personnel
4.1.2.3 Management Representative
4.1.3 Management Review

4.2 Quality System

4.3 Contract Review

4.4 Design Control

4.4.1 General
4.4.2 Design and Development Planning
4.4.2.1 Activity Assignment
4.4.2.2 Organizational and Technical Interfaces
4.4.3 Design Input
4.4.4 Design Output
4.4.5 Design Verification
4.4.6 Design Changes

4.5 Document Control

4.5.1 Document Approval and Issue
4.5.2 Document Changes/Modifications

4.6 Purchasing

4.6.1 General
4.6.2 Assessment of Sub-Contractors
4.6.3 Purchasing Data
4.6.4 Verification of Purchased Product

4.7 Purchaser Supplied Product

4.8 Product Identification and Traceability

4.9 Process Control

4.9.1 General
4.9.2 Special Processes

4.10 Inspection and Testing

4.10.1 Receiving Inspection and Testing
4.10.2 In-Process Inspection and Testing
4.10.3 Final Inspection and Testing
4.10.4 Inspection and Test Records

4.11 Inspection, Measuring, and Test Equipment
(Contains several subparagraphs a-j and includes hardware and software)

4.12 Inspection and Test Status

4.13 Control of Nonconforming Product

4.13.1 Nonconforming Review and Disposition

4.14 Corrective Action

4.15 Handling, Storage, Packaging, and Delivery

4.15.1 General
4.15.2 Handling
4.15.3 Storage
4.15.4 Packaging
4.15.5 Delivery

4.16 Quality Records

4.17 Internal Quality Audits

4.18 Training

4.19 Servicing

4.20 Statistical Techniques

---

should register five plants under one certificate, all five plants run the risk of losing their registration if anyone of the other four plants should be nonconforming.

Some Clarifications Regarding the ISO 9000 Series

Despite the increasing number of seminars, articles and speeches about the ISO 9000 series, confusion still persists. One the most often asked questions about the ISO series is: "How does it apply to my industry?" Indeed, although each of the three standards do not currently exceed seven pages in length, their contents—intended for *all* industries—is open to a wide variety of interpretations. In an attempt to clarify or otherwise bring particular relevance to the standard, several federal and private agencies have recently adopted the ISO 9000 series (essentially 9001) by *aligning* its contents with industry specific regulations. For example, the U.S. Food and Drug Administration (FDA) has already incorporated the ISO 9001 series into its Good Manufacturing Practices (GMP). Similarly, the American Petroleum Industry (API) is also aligning its Q1 document with ISO 9001. The Department of Defense is about to replace its MIL-Q-9858A and MIL-I-45208A quality system standards with the ISO 9001 standards. Other federal agencies seriously considering adopting the ISO 9000 include the Federal Aviation Administration (FAA) and the National Aeronautics and Space Administration (NASA).

In the U.K., the Department of Trade and Industry, in association with the British Computer Society, developed a software quality management initiative known as TickIT. A flyer advertising TickIT states that "[T]he aim of TickIT is to achieve improvements in the quality of software products and information systems throughout the whole field of Information Technology supply including in-house development work." [1](See also Chapter 9).

To assist the process industry, the European Chemical Industry Council (CEFIC) published in July 1991 its EN 29001/ISO 9001 *Guidelines for Use By the Chemical Industry*. In addition, various European councils are developing industry specific standards. The EN 46001 document developed for the medical device sector and the EN 2000 and EN 3042 for aerospace products are but a few examples.[2]

---

[1] TickIT is not an international standard and should not be interpreted as such. It is one more British standard.

[2] *Europe: A-Report*, Winter 1991-1992, p. 6. Published by the U.S. Department of Commerce International Trade Administration.

Commercial laboratories looking for ISO registration should follow the ISO/IEC Guide 25 "General Requirements for the Competence of Testing Laboratories," rather than the ISO 9003 standard.[1] Failure to do so may lead to some potential difficulties, particularly if test results are to be recognized/accepted by some EC member states. In the United Kingdom, for example, NAMAS (the UK national laboratory accreditation system) requests that in order to be fully accredited, commercial testing services will have to be accredited by NAMAS and no other registrar. Other countries have adopted a similar policy. American commercial laboratories should contact the American Association for Laboratory Accreditation (A2LA), for further information.[2]

The EC's Technical Harmonization Directives Regarding Regulated Products

With U.S. exports to the European Community (EC) exceeding $100 billion in 1991, the assurance that the EC's new testing and certification program would not restrict U.S. exports is of great importance to U.S. exporters. The Single Internal Market, which took effect on January 1, 1993 and has lead to the formulation of 282 White Paper directives, is of significant importance to American businesses (Table 1.6). Indeed, U.S. firms will need to know what directives apply to their products and what the status of implementation is for that legislation.

The European Community's directives are subdivided into two major groups: "L" documents and "C" documents. The "L" series of the *Official Journal*, of which they are currently 282, "lists directives, recommendations and decisions of the European Community that have been approved by member states through the EC council and are now

---

[1] See Appendix F for a list of ISO/IEC Guides.

[2] John W. Locke, "Quality Standards for Testing Laboratories," American Association for Laboratory Accreditation, 656 Quince Orchard Road #304, Gaithersburg, MD 20878-1409

binding."[1] The "C" series, which currently consists of 320 directives, are proposed directives that are being considered by EC governments.

Table 1.6. Member State Implementation of Single Market Directives*

	Implemented	Not Implemented	Not Applicable
Germany	103	28	6
Belgium	96	35	6
Denmark	123	7	7
Spain	105	25	4
France	119	13	5
Great Britain	113	17	6
Greece	104	24	5
Italy	73	59	5
Ireland	95	35	6
Luxembourg	90	39	8
Netherlands	99	32	6
Portugal	109	23	2

* Source: *Business America*, February 24, 1992

Of particular importance to American businessmen are the technical harmonization directives first made official on December 13 1990.[2]

Technical Harmonization Directives—General Guidelines

The general guidelines consist of fourteen steps (a-n). An *overview* of these guidelines is presented here (all quotations are from the *Official Journal of the European Communities* cited below, emphasis added).

[1] "List of European Community 1992 Directives and Proposals," Single Internal Market Information Service, July 4, 1991.

[2] Council Decision of 13 December 1990 concerning the modules for the various phases of the conformity assessment procedures which are intended to be used in the technical harmonization directives. *Official Journal of the European Communities* (31.12.90) No I. 380/13 (emphasis added).

(a) "the essential objective of a conformity assessment procedure is to enable the public authorities to ensure that products placed on the market conform to the requirements as expressed in the provisions of the directives, *in particular to the health and safety of users and consumers;"*

(b) "conformity assessment can be subdivided into modules which relate to the design phase of products and to their production phase;" (see Table 1.7 below)

(c) "as a general rule a product should be subject to both phases before being able to be placed on the market if the results are positive;" (specific directives may provide for different arrangements)

(d) "there are a variety of modules which cover the two phases in a variety of ways. The directives shall set the range of possible choices . . . to give the public authorities the high level of safety they seek, for a given product or product sector;"

(e) The council recognizes that "the directives will take into consideration, in particular, such issues as the appropriateness of the modules to the type of products, the nature of the risk involved, the economic infrastructure of the given sector, etc.;"

(f) "the directives will, in setting the range of possible modules for a given product or product sector, attempt to leave as wide a choice to the manufacturer as is consistent with ensuring compliance with the requirements;"

(g) "the directives should avoid imposing unnecessary modules which would be too onerous relative to the objectives of the directive concerned;"

(h) "notified bodies should be encouraged to apply the modules without unnecessary burden for the economic operators . . . ;" [**Note:**

Notified bodies include laboratories, certification bodies (i.e., registrars which audit quality systems) or inspection bodies. Notified bodies perform third party audits and are accredited according to the EN 45000 series by various national accreditation councils such as, for example, the U.K.'s National Accreditation Council for Certification Bodies (NACCB) or the Dutch Raad voor Certificatie (RvC), and France's Association Française Assurance Qualité (AFAQ).]

(i) "in order to protect the manufacturers, the technical documentation provided to notified bodies has to be limited to that which is required solely for the purpose of assessment of conformity;"

(j) "whenever directives provide the manufacturer with the possibility of using modules based on quality assurance techniques, the manufacturer *must be able to have recourse to a combination of modules not using quality assurance, and vice versa*, except where compliance with the requirements laid down by the directives requires the exclusive application of a certain procedure."

(k-n) These paragraphs relate to notified bodies and basically state that a member state shall rely on notified bodies which have "the technical qualifications required by the directives." In addition, these notified bodies must conform with the EN 45000 standards for notified bodies. Finally, a list of notified bodies will be published (hopefully sometime in 1993), by the Commission in the *Official Journal of the European Communities.*

Having read the above guidelines, the reader may well ask the following two important questions: "Do I manufacture product(s) or provide a service that fall(s) under one (*or more*) of the EC regulated product directives?" and if so, "Which module(s) must I follow?"

European Directives

The 282 directives affect a wide range of products and services (see Table 1.7)[1]. However, despite the EC Commission's diligent work, only a handful of products or product sectors directives have been completed to date.

Since the adoption of the Resolution of May 7, 1985 concerning a new approach to technical harmonization and standardization, the Council has adopted nine directives which affect the following product types: *toys, construction products, pressure vessels, electromagnetic compatibility machinery, personal protective equipment, gas appliances, non-automatic weighing instruments, medical devices and telecommunications terminal equipment.* If your company exports any of the above regulated products to the EC, it will be subject to one of the eight modules, which in turn determine rules for affixing the CE mark.

Certification Options (Modules) and the CE Mark

Products falling under the scope of a directive will have to carry a "CE" mark. The CE mark "is intended above all for the market inspectors in the Member States and as such does not claim to be a mark of quality, safety or environmental protection as generally understood by consumers. It is therefore intended for the trade and its aim is not to provide an encyclopedia of technical information about the product since, if the mark appears on a product, the inspector must consider the product to be in conformity."[2] Since the meaning of the CE mark differs from one directive to another, an exporter must first know

---

1 Timothy M. Devinney and William C. Hightower, *European Markets after 1992* (Lexington, MA: Lexington Books, 1991), provide a detailed list of all directives and their status in Table B1: European Community Directives, Decisions, Proposals, and Regulations, pp. 223-279.

2 *Official Journal of the European Communities*, op. cit., p. 5.

Table 1.7. Product Sector Directives

(for further details see *Business America*, February 24, 1992, p. 25)

Subject Area	Sample of Affected Products
In standards, testing and certification.	Harmonization of standards for: simple pressure vessels, toys, construction products, machine safety, agricultural and forestry tractors, cosmetics, quick frozen foods, flavorings, food emulsifiers and preservatives, gas appliances, weighing instruments, radio interferences, lawn mowers (noise), household appliance (noise), detergents, and other products including infant formula.
New Rules for harmonizing packing, labelling and processing requirements	Irradiation, nutritional labelling, extraction solvents, ingredients for food and beverages, etc.
Harmonization of regulations for the health industry	Medical specialties, high technology medicines, veterinary medicinal products, active implantable medical devices, pharmaceuticals, medical devices, in-vitro diagnostics
Changes in government procurement regulations	
Harmonization of regulation of services	Tourism, electronic payment cards, information services, road haulage, banking, railways, etc.
Liberalization of capital movements	
Consumer protection regulations	
Harmonization of taxation	
Harmonization of laws regulating company behavior	Copyrights, bankruptcy, trademark, etc.
Harmonization of veterinary and phytosanitary controls	
Elimination and simplification of national and transit documents and procedures for intra-EC trade	Elimination of customs formalities and the introduction of common border posts, etc.
Harmonization of rules pertaining to the free movement of labor and the professions within the EC	Mutual recognition of higher educational diplomas, training of engineers, etc.
Autos	Speed limitation devices, braking devices, safety belts and restraint systems, tires, indicator lamps, etc.
Health and safety/social charter	Many items
Environment	PCBs, asbestos, shipment of hazardous waste, etc.
Telecommunications	Type approval of terminal equipment, telecom services with open network provisions, computer programs, public digital telecommunication networks, etc.
Energy	

which directive and hence which modules applies to his product.

The eight modules are:

Module A: Manufacturer self declaration of conformity
(Module Aa: An extension of Module A)
Module B: EC type examination
Module C: EC declaration of conformity to type
Module D: Production quality assurance (ISO/EN (2)9002)
Module E: Final inspection and testing (ISO/EN (2)9003)
Module F: Product verification by EC third party series production.
Module G: Same as F but for unit verification
Module H: Full quality assurance (ISO/EN(2)9001)

The general requirement of each module is summarized in Table 1.8. With respect to the above products for which directives have already been written, the following modules apply:

Toys - Manufacturer self-declaration (Module A)
Construction products - at a minimum, manufacturer registration of production quality assurance system (Module D).
Pressure Vessels - EC type examination (Module B)
Electromagnetic Compatibility - Manufacturer self declaration (Module A)
Machinery - Manufacturer self-declaration (Module A)
Personal protective equipment - EC type examination (Module B) with quality control system registration for higher risk equipment (Modules D, E, F or H)
Gas Appliance - EC type examination AND either quality assurance system registration OR on-site checks of appliances (Modules B plus C, D, F or H)
Non-automatic weighing instruments - EC type examination AND quality assurance registration or EC verification (Module B, C-H)

Medical devices - various options depending on risk level ranging up to Module H.

Telecommunications terminal equipment - EC type examination OR declaration of conformity with full quality assurance (Module H).

For a brief review of telecommunications, medical and food and beverages directives as well as useful addresses on how to obtain further information regarding directives, the reader is referred to Appendix A. Manufacturers of electronic components should refer to Appendix H. Electronic Components

Once an ISO 9001, 9002 or 9003 quality system has been implemented and products conform to one or more of the technical harmonization modules (A-H), what must an exporter do? Although there are many options available to an exporter (see Table 1.8), all modules (except for module A) require the participation of a notified body to either test or check a product according to specified statistical plans **or** *approve* the supplier's quality system. Product verification will likely be performed by laboratories or inspection bodies which have the required technical expertise (see Chapter 9). The approval/verification of quality systems will continue to be performed by accredited registrars which will conduct periodic third party audits (i.e., contracted by the suppliers). A partial list of U.S. and European registrars including their accreditation status can be found in Appendix C.[1]

---

[1] For more information on the ISO 9000 registration process the reader is referred to James Lamprecht's *ISO 9000: Preparing for Registration*. Marcel Dekker, N.Y., N.Y. 1992. See also James Lamprecht, "ISO 9000 Implementation Strategies," in *Quality*, November 1991, pp. 14-17.

Table 1.8 Conformity Assessment Modules

	A. Internal Control of Production	G. Unit Verification	H. Full Quality Assurance
DESIGN	Manufacturer Keeps technical documentation at the disposal of national authorities Aa. Intervention of notified body	Manufacturer - Submits technical documentation	EN 29001 ISO 9001 Manufacturer - Operates an approved quality system (QS) for design Notified body - Carries out surveillance of the QS - Verified conformity of the design - Issues EC design examination certificate
PRODUCTION	A. Manufacturer - Declares conformity with essential requirements - Affixes the CE mark Aa. Notified body - Tests on specific aspects of the product - Product checks at random intervals	Manufacturer - Submits product - Declares conformity - Affixes the CE mark Notified body - Verifies conformity with essential requirements - Issues certificate of conformity	Manufacturer - Operates an approved QS for production and testing -Declares conformity - Affixes the CE mark Notified body - Carries out surveillance of the QS

Table 1.8 Continued

B. Type Examination			
Manufacturer submits to notified body - Technical Documentation - Type Notified Body - Ascertains conformity with essential requirements - Carries out tests, if necessary - Issues EC type-examination certificate			
C. Conformity to Type	D. Production Quality Assurance	E. Product Quality Assurance	F. Product Verification
Manufacturer - Declares conformity with approved type - Affixes the CE mark	EN 29002/ISO 9002 Manufacturer - Operates an approved quality system (QS) for production and testing - Declares conformity with approved type - Affixes the CE mark	EN 29003/ISO 9003 Manufacturer - Operates an approved quality system (QS) for inspection and testing - Declares conformity with approved type, or the essential requirements - Affixes the CE mark	Manufacturer - Declares conformity with approved type, or with essential requirements - Affixes the CE mark
Notified body - Tests on specific aspects of the product - Product checks at random intervals	Notified body - Aproves the QS - Carries out surveillance of the QS	Notified body - Approves the QS - Carries out surveillance of the QS	Notified body - Verifies conformity - Issues certificate at conformity

Source: *Official Journal of the European Community (31.12.90)*, p. 26.

Summary and Conclusions

In view of the great confusion surrounding EC directives and ISO 9000 registration, it is important to summarize the following key points:

• *In most cases, EC directives set a range of possible procedures to allow the manufacturer a choice of how to demonstrate his product's conformity to legal requirements. For example, product certification under the EC directive on gas appliances can be accomplished through*

EC type examination, accompanied by any one of modules C through F. Telecom terminal equipment can be certified by EC type examination, accompanied by either modules C or D, or by full quality assurance procedures (module H).

- *Manufacturer self-declaration of conformity to EC requirements is permitted for machinery (except for high risk equipment), toys, electromagnetic compatibility and weighing instruments for commercial use, as well as for certain classes of personal protective equipment, pressure vessels and equipment, medical devices, and recreation craft.*

- *ISO 9000 is not a blanket requirement for all regulated products in the EC.*

- *EC legislation mandates some type of third-party involvement in product certification for regulated products. EC member states are responsible for determining the competence of test labs and certification bodies that apply for recognition under the EC system.* ***Member states*** *notify the EC commission of their selection—thus the term "notified" bodies—by task and by directives.*

- *Notified bodies (which are likely to include registrars such as BSI and Lloyds (U.K.), TUV (Germany), DNV (Norway), AES (Apave, France) and U.S. registrars) can subcontract specific activities as long as subcontractors comply with EN 45000 (See James L. Lamprecht, ISO 9000: Preparing for Registration, Chapter 11).*

- *Under the EC system, member states can designate notified bodies only from within the EC; notified body status does not extend to subsidiaries or related enterprises located in third country. However, the system does permit authorization of entities located in a third country to perform third-party certification, accreditations or approvals provided that these entities operate under the mutual recognition agreement negotiated between the EC and government authorities of that country. For more specific references regarding notified bodies negotiations with third countries, see Appendix G.*

• *According to the draft regulation, the manufacturer, or his authorized representative in the EC, is responsible for affixing the CE mark to the apparatus or to the packaging, the instructions for use or the guarantee certificate.*[1]

Having reviewed some of the complex regulations and directives affecting the European Community's regulated products and their relationship with the EN 29000/ISO 9000 series, we now turn our attention to an analysis of the ISO 9000 series as a quality assurance system.

---

[1] Italicized text quoted from "EC Testing and Certification Procedures Under the Internal Market Program," April 1, 1992. United States Department of Commerce, International Administration, pp. 4 *et passim.*

2 The ISO 9000 Series as a System

Explaining how to document a quality assurance system based on the International Organization for Standardization (ISO) 9000 series can bring about lively debates, arguments and counter arguments as to what is the proper philosophical construct to adopt. After reading the standard, it is not uncommon for some seminar participants to observe, with some disappointment: "Is that all there is to ISO? We do much more at our plant!" Naturally, suppliers to the nuclear, aerospace or pharmaceutical industries interpret and perceive the ISO 9000 standards from a very different point of view than a manufacturer of cardboards or bottle caps, for example.[1] Nevertheless, irrespective of the industry, one of the questions that usually surfaces during such discussions concerns the purpose or objective(s) of "ISO implementation." Some would proselytize that the ISO series of standards can only be implemented within the framework of a Total Quality philosophy.[2] In effect, that faction sees the ISO series as providing a structure to the very tenets of total quality. To others, the ISO series is nothing more than a baseline model/foundation upon which one can build a total quality superstructure. Others, would argue that the ISO series has little to do with quality and is indeed a step backwards. Extremists in that latter group believe that the ISO series might in fact represent the antithesis to total quality.

I would propose that unless a definition of total quality is first agreed upon by *all* parties (perhaps a "mission impossible" objective), it is essentially impossible to say who is right or wrong. Even then, the consensus might be that all are partly correct in their assessment.

---

1 The ANSI/ASQC Q90-1987 standard does recognize the need for variability: "The quality system of an organization is influenced by the objectives of the organization, by the product or service, and by the practices specific to the organization, and therefore, the quality system varies from one organization to another." ANSI/ASQC Q90-1987, p. 1 paragraph 0.0 INTRODUCTION.

2 I am avoiding the use of the expression Total Quality Management (TQM) for I have often wondered if the expression TQM might not be an oxymoron. I believe the use of company-wide quality (or some similar expression) to be more meaningful. See also Gilbert Fuchsberg's "'Total Quality' is Termed Only Partial Success," in *Wall Street Journal* of Thursday October 1, 1992 p. B8.

Indeed, opinions as to what the ISO 9000 series is or is not will depend as to where you are coming from, how far you have traveled (along the tortuous road of quality), and for how long. Nonetheless, irrespective of first impressions, the overwhelming majority of participants do believe that the ISO series is a good baseline system.

What Do You Mean by Total Quality?

At the risk of opening several Pandora boxes I will attempt to answer this elusive question. When asked to define what is meant by total quality, most people would cite:

1. Employee participation (Ishikawa)[1]
2. Continuously improving all processes (Juran's spiral of continuous improvement)
3. Monitoring your processes using appropriate statistical techniques (Deming)
4. Surveying your customers and benchmarking your competitors (Malcolm Baldrige)
5. Innovation, in order to remain competitive (T. Peters)

Although the reader would be hard pressed to find specific "ISO references" to any of the above five points (except point two which is covered in paragraphs 4.14 and 4.13 in ISO 9001 and ISO 9002 respectively), none of the above five points - with the possible exception of point five (see Chapter 10) - contradicts the intent of the ISO series of standards.[2] This becomes obvious when one realizes that the ISO 9000 series provides a *model* for quality *systems*. ISO 9001 (ANSI/ASQC Q9001-1987) for example, is entitled *Quality Systems --*

---

[1] I have included in parentheses the name of the most famous proponent of the proposed concept. I do not mean to imply or suggest that individuals such as Ishikawa, Juran, Deming, Peters and, indeed, Feigenbaum (who coined the expression "Total Quality Management") are solely known for one and only one idea. Deming, for example, is well known for his famous fourteen points.
[2] With regard to total quality and the implementation of ISO, see also pages 18-20 of my *ISO 9000: Preparing for Registration* (New York: Marcel Dekker and ASQC Quality Press, 1992).

Model for Quality Assurance in Design/Development, Production, Installation, and Servicing.

The ISO 9000 Series as a System

> **Quality System**: The organizational structure, responsibilities, procedures, processes, and resources for implementing quality management.[1]

The above quotation still leaves us with the question, what is meant by a system? A system is a "group of *interacting* items that form a unified whole or that are under the influence of forces in some *relationship*".[2] Perusing the ISO 9001, 9002 or 9003 quality assurance systems, one soon notices that the interacting items are the various clauses which form a unified whole embodied by the documented and operating quality assurance model itself (see Figure 2.1).

Within any organization, numerous functional/departmental relationships operate to make the system work. As Parsegian explains,

> Most systems that have significance to our world are likely to have controls and to be "designed" to use feedback influences through which to exercise control. To serve as a *control system*, in turn, implies the existence of a plan, or *design*; or a *purpose*, or objective.[3]

---

[1] ANSI/ASQC Q90-1987. *Quality Management and Quality Assurance Standards - Guidelines for Selection and Use*, p. 2.
[2] V. L. Parseghian. *This Cybernetic World of Men, Machines, and Earth Systems* (1973). Anchor Books, p. 23.
[3] Parseghian, op. cit. p. 24.

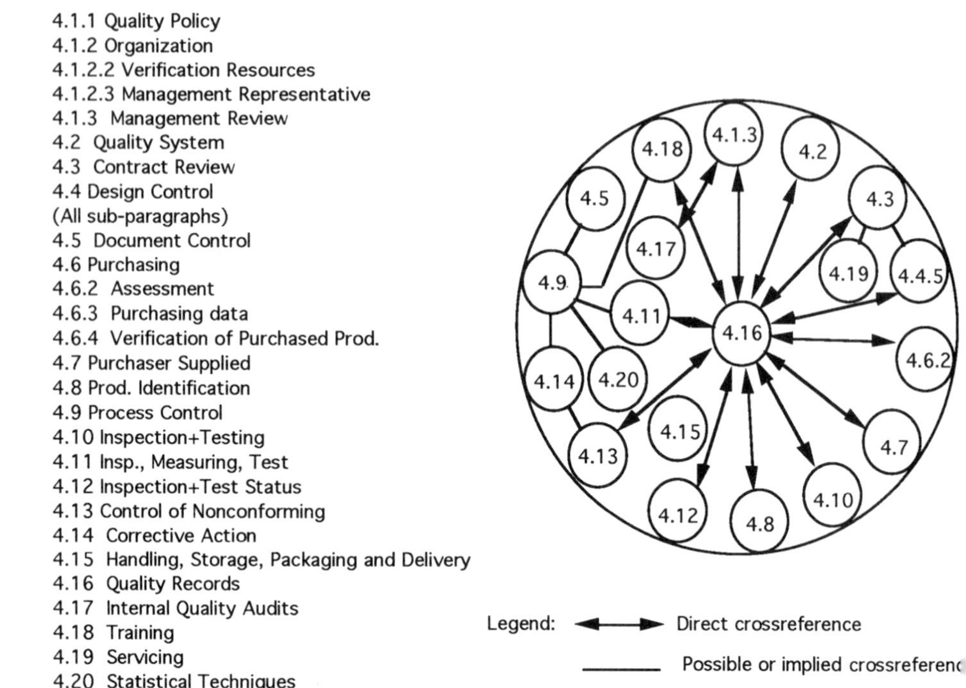

Figure 2.1: Latent and Actual Interrelationships for Some ISO 9001 Clauses

The above quote is particularly significant when placed within the context of "ISO implementation." Companies which have struggled with the implementation of an ISO 9000 type quality assurance system, have invariably either failed to establish their *purpose* for wanting to achieve registration, neglected to design a plan outlining how they will achieve "ISO readiness," or both.

What to Consider When Implementing an ISO 9001, 2 or 3 Quality System

Anyone who has read one of the ISO standards will rarely, if ever, contest their contents. Most people agree that the standards provide a basic framework which can be adapted to each company's individual needs and requirements. However, when the time comes to actually ensure that what should be done is actually done, confusion seems to reign. The standard's prose, written to apply to just about anyone, becomes very murky. What was obvious and logical a short while ago becomes confusing. Questions multiply at an exponential rate: "What do they mean by process control?", "Must I calibrate all of my instruments?", "How do you define contract review?", "What is a special process?". Whereas it is true that reference to the ISO 9004 management guidelines does provide some valuable information, in some cases, referral to the guidelines leads to more questions.

To facilitate implementation, it is helpful to subdivide the standards' clauses into two major categories: a) clauses that focus on the management/monitoring of the quality assurance system and are thus likely to apply throughout a company and b) department-specific or special clauses.

For example, with respect to the ISO 9001 standard, companywide clauses could include:

- 4.1 Management Responsibility

  4.1.2.1 Responsibility and Authority
  4.1.2.2 Verification Resources and Personnel
  4.1.2.3 Management Representative
  4.1.3 Management Review

- 4.2 Quality System
- 4.3 Contract Review (could be specific to one department)

- 4.5 Document Control
- 4.16 Quality Records
- 4.17 Internal Quality Audits

Department specific or special clauses could include everything else:

- 4.4 Design Control

    4.4.1 General
    4.4.2 Design and Development Planning
    4.4.2.1 Activity Assignment
    4.4.2.2 Organizational and Technical Interfaces
    4.4.3 Design Input
    4.4.4 Design Output
    4.4.5 Design Verification
    4.4.6 Design Changes

- 4.6 Purchasing

    4.6.1 General
    4.6.2 Assessment of Sub-Contractors
    4.6.3 Purchasing Data

- 4.7 Purchaser Supplied Product
- 4.8 Product Identification and Traceability
- 4.9 Process Control
- 4.10 Inspection and Testing
- 4.11 Inspection, Measuring, and Test Equipment
- 4.12 Inspection and Test Status
- 4.13 Control of Nonconforming Product
- 4.14 Nonconformity Review and Disposition
- 4.15 Handling, Storage, Packaging and Delivery
- 4.18 Training (could be decentralized by department)
- 4.19 Servicing
- 4.20 Statistical Techniques

The primary challenge is to identify which department(s) is responsible for which clause(s). This is easier said than done because,

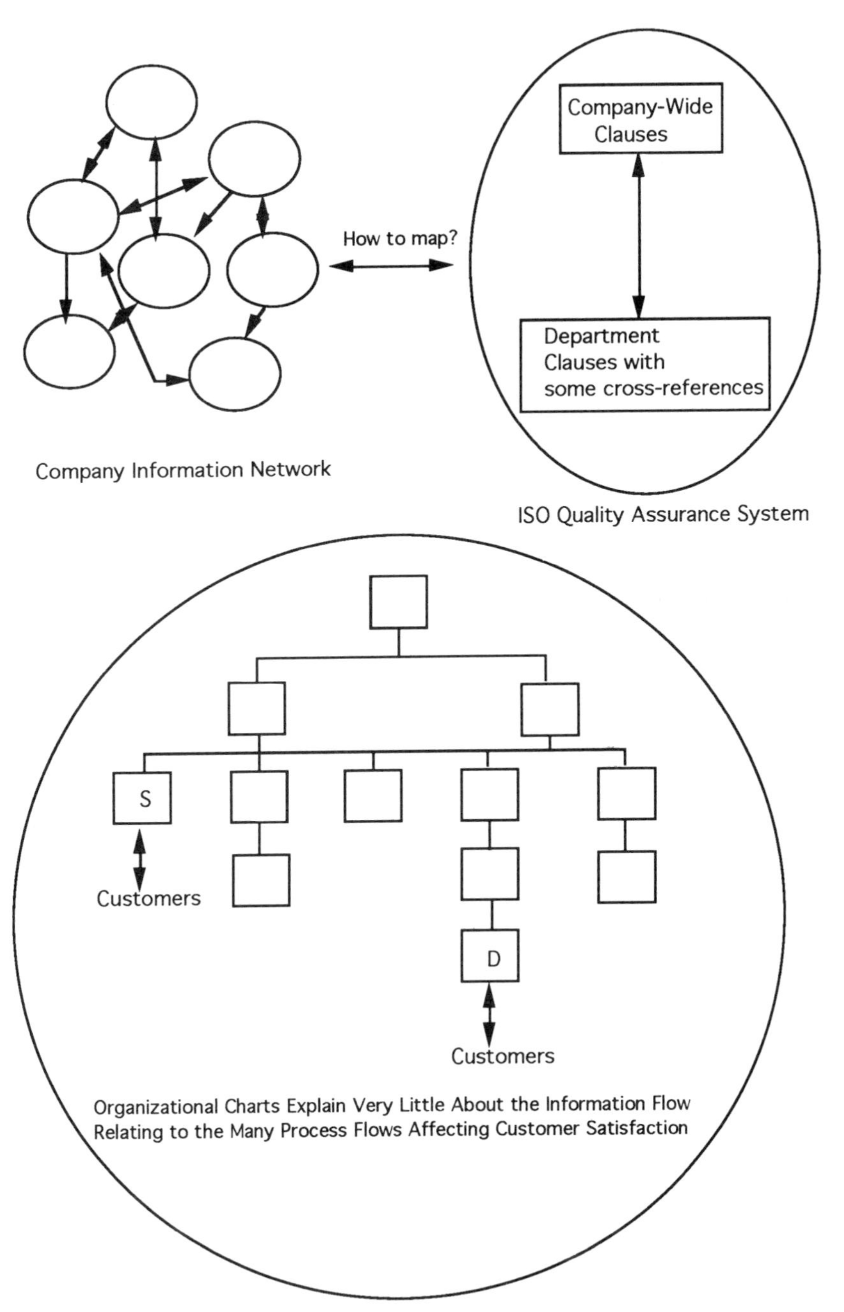

Figure 2.2: Matching ISO's Quality Assurance System to a Company's Information Network

in most cases, the interrelationship between ISO clauses and department is multifunctional. Moreover, the inherent organizational structure of the standard is not always conducive to an easy alignment with a company's informational flow (see Figure 2.2).[1] Consequently, to ensure as smooth an implementation process as possible, it is important to consider the *cybernetics* of ISO 9000 implementation.

The Cybernetics of ISO Implementation

When a company embarks on the road to ISO registration it rarely, if ever, does so from ground zero. In most cases, the company in question will have a quality assurance system in place. What is certain to vary, however, is the complexity, efficacity and effectiveness of the Q.A. system which itself is a function of several characteristics including, among other things, the type of industry, level of government regulation (if any), volume of production, etc. Thus, whereas it is true that some highly regulated industries (medical, pharmaceutical, avionics/aerospace, suppliers to the military, etc.) are likely to find that little adjustments will need to be made to their systems, others will find the task significantly burdensome but, hopefully, rewarding.[2]

In the majority of cases, the potential customer to the ISO 9000 series will find that although the company's Q.A. system addresses the majority of activities listed within the ISO standard, the Q.A. system could use some revamping and may even be bordering on chaos.

It is the nature of any system (Q.A. included) to continuously deteriorate through time. This state of deterioration towards which systems tend to stabilize is known as entropy maximization. Based on

---

1 The use of organizational charts can be helpful, see for example James Lamprecht, *ISO 9000: Preparing for Registration* (New York: Marcel Dekker and ASQC Quality Press, 1992), Chapter 6 (figure 6.1) and Chapter 9.

2 Aerospace suppliers have long been exposed to the Coordinated Aerospace Supplier Evaluation (CASE), an informal association of aerospace prime contractors.

the current disorganized state of my desk I would have to conclude that this particular system has achieved a high level of entropy. In order to bring about order and prevent further increase in entropy, I would have to spend some time and energy re-organizing the chaotic information content of my desk. A similar analogy applies to Q.A. systems, particularly if they are not monitored for long period of times. Indeed, quality assurance systems which are not monitored (audited) periodically will exhibit signs of chaos and/or malfunction. To bring about or regain order, one must input information and energy back into the system—this process is diagramed in Figure 2.3. This is usually achieved via documentation, document control, training or other activities such as internal quality audits (see Chapter 6). Not surprisingly, these activities, required by paragraphs 4.2, 4.5, 4.18, 4.17 (and others) of ISO 9001, tend to be the most difficult to implement.

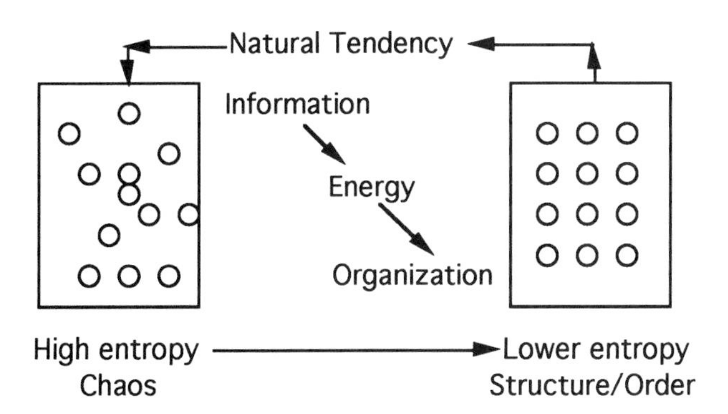

Information -->Documentation/Document Control

Energy --> Internal audits to maintain system organized and prevent deterioration back to chaos.

Figure 2.3: The Organization of a Quality Assurance System

How to Bring About Order Consistent with the "Show Me Mapping"

Introducing an ISO 9000 type quality assurance program within a company is not an easy task particularly if the company already has a quality assurance program in place. In most cases, the quality assurance system is either structured to please one major customer or is an amalgamation of several quality assurance plans aimed at attempting to satisfy the occasional conflicting requirements of many customers. As far as many suppliers are concerned, the ISO 9000 series is nothing more than an additional set of standards to an already long list of standards: Mil Q 9858A (and many other Mil standards), Boeing's D1-9000, Airbus' AIQI (Airbus Industries Quality Instruction), FAR, FDA/GMP, Ford Q-101, API, IEEE, ASME, etc. Faced with a plethora of standards and regulations, one can sympathize with the pragmatic business person who, when requested to "look into ISO", simply asks "Why?"

Naturally, if an extensive quality assurance system is already in place, the logical and acceptable temptation will be to try to incorporate the ISO 9000 series within the current system. For some companies, already used to having several quality plans designed to "handle" their customers' idiosyncracies, the task may well be routine. For others, however, the experience may well induce trauma. The good news is that ISO requirements should already be included in most if not all of your customers' quality assurance demands. Consequently, in theory, little additional effort should be required to adapt one's quality assurance plan to the ISO standard. In practice, however, events might be different simply because one of the main focuses of the ISO registration process is to *verify that you are indeed doing what is claimed in your quality manual.* This approach may perhaps not come as a surprise to readers who are accustomed to having their documented Q.A. system audited. It may however surprise others who may have assumed that achieving ISO 9000 registration merely consisted of hiring a consulting firm to devise and document an impressive Q.A. system which may, at best, vaguely resemble reality.

The best strategy to adopt when embarking on the road to ISO registration is to adopt a simple model: *design a quality assurance model from the bottom up to ensure that what is done is indeed what is documented.* This advice, trite as it may be, is perhaps the best suggestion I can offer. It is indeed surprising to find out how many quality assurance systems are written to describe what *should* be done (or was *once* done), rather than what is *currently done.*

To understand how to approach the ISO 9000 series, one must first understand the "Show Me Mapping" process favored by most registrars conducting *third-party* audits. The process, outlined in Figure 2.4, emphasizes not only conformance to the various ISO clauses, but also conformance to how accurately the supplier's (i.e., the entity wishing to achieve ISO registration), claims are documented and practiced. This dual objective of conforming to an interpretation of the standard and ensuring that what is written down is indeed practiced is not easily achieved.

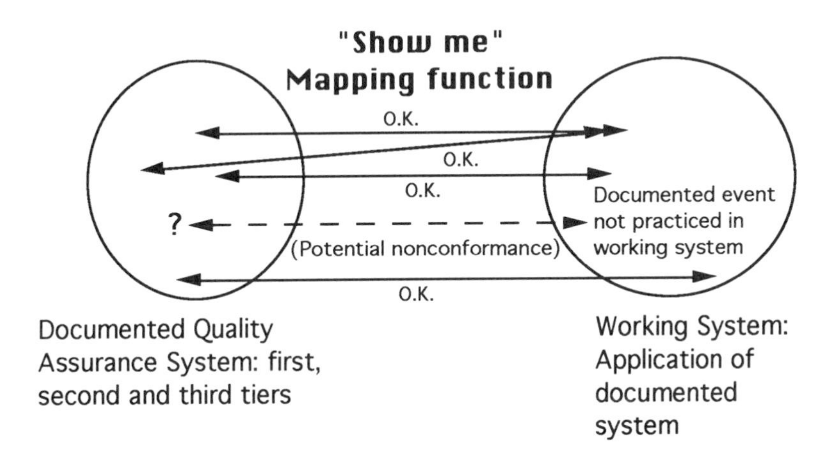

Figure 2.4 Raising Nonconformances

One of the easiest ways to achieve ISO compliance is to focus on the following steps:[1]

1. Familiarization with the standard
2. Assessment of how the current Q.A. system addresses ISO requirements
3. Implementation of corrective actions where the two systems do not match
4. Continuous monitoring via internal quality audits to ensure effectiveness of the documented system

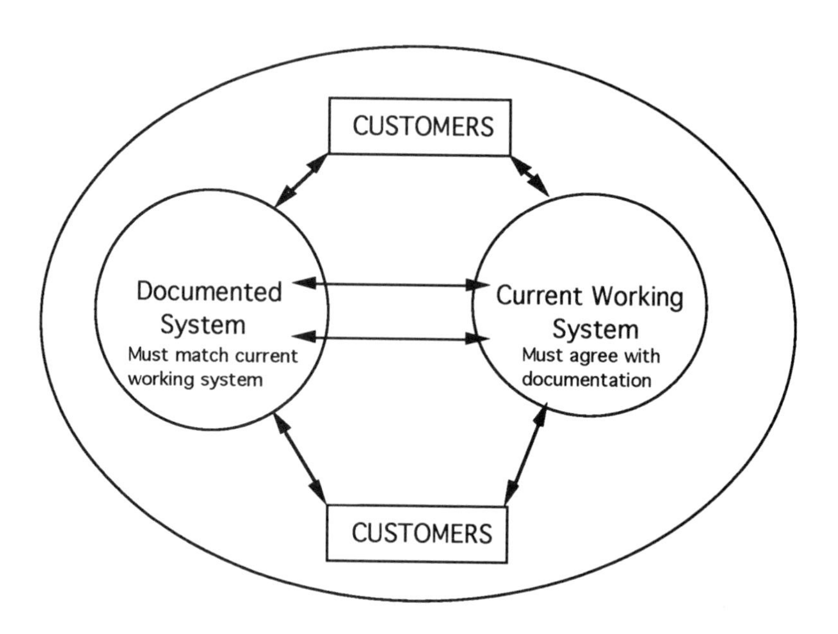

Figure 2.5 ISO Implementation: Do NOT Forget Your Customers

---

[1] See the author's *ISO 9000: Preparing for Registration*, op. cit. for further details.

The quality assurance system should: (1) suit the supplier's need, (2) not be so restrictive as to become impractical and thus ineffective and (3) should be continuously improved. Whatever the final format, the Q.A. system should never ignore its ultimate purpose: *to bring about customer satisfaction* (see Figure 2.5). Indeed, the implicit driving force behind every registration effort should not be the registrar nor the ISO certificate, but rather the formulation of a well thought out effective system designed to bring about improved performance.

Summary and Conclusions

The ISO 9000 standards should not be perceived as a panacea to quality. Rather, it is a baseline model for quality assurance which can be upgraded/improved as you see fit. For example, whereas it is true that the standards do not currently have clauses on the *cost of quality* and process or equipment *reliability*, nothing prevents the user to go beyond the standard's intent and include these requirements within the company's quality assurance *system*.

As a system, the ISO quality assurance model can only survive as an operational system *if and only if* various departments continuously *interact* to monitor/audit the system and its effectiveness. To prevent, or rather, contain the natural tendency of any system to deteriorate to maximum entropy, and thus chaos, the system must continuously be monitored. Having presented some of the global aspects of the ISO 9000 series, let us now focus on the interpretation of some specific clauses.

3 How Do You Interpret This Paragraph?

Introduction

Given the vast number of companies which have applied and will continue to apply for ISO 9000 registration, it is not surprising to find out that there are as many interpretations of ISO 9000 as there are readers of the standards. Since it is impossible to present an interpretation of all twenty ISO 9001 paragraphs from all possible perspectives, I will limit myself to offering a few examples which I hope will suffice to explain how one should document a quality assurance system which will conform to the ISO 9000 series of standards. Clauses not reviewed in this chapter are reviewed in Appendix A. Having lectured on the subject of the ISO 9000 series over the past three years to well over 700 participants, I have been fortunate to collect a vast array of examples derived from thousands of questions and hundreds of discussions. The following account is but an outline designed to emphasize one premise: you must learn to adapt the ISO 9000 series to your every-day operational needs while at the same time satisfy its requirements.

Does This Paragraph Apply to My Industry?

Anyone who has read the ISO 9000 series of documents quickly realizes that much time is spent interpreting the content. This is most vividly revealed during public seminars. When teams of three to five individuals from various industries, are asked to review and comment on one or two ISO 9001 paragraphs, the team reporter invariably concludes that although each team member usually addresses the paragraph's content, the company they represented approached each ISO clause in a slightly different manner. When the audience is then asked why there is such variance in interpretation, the unanimous answer is "It depends on our customers' requirements." Indeed, it is clear that the type of industry within which a company operates invariably dictates the type of customer and/or regulatory (government or private) agencies it must abide to.

This in turn is reflected in how well the quality assurance system of a particular company matches the ISO "requirements." Thus, before focusing on how to implement an ISO quality assurance system, it is important to first know where along the industrial continuum a company stands.

A Typology of ISO 9000 Applicants

The myriad of companies applying for ISO 9000 registration is mind boggling. Given the broad diversity of applicants and the purposefully all encompassing prose of the ISO 9000 series of standards, it is imperative to recognize that a monolithic interpretation of the standard cannot be universally applied to all companies. Hence, the enlightened recognition by members of the TC/176 committee that the standards may on occasion be "tailored for specific contractual situations" (ANSI/ASQC Q9001, p.1 paragraph 0.0 Introduction).

Over the past thirty years, organizational psychologists have attempted to broadly classify organizations based on the type of production and processes. Building on research conducted in the 1960s, particularly Joan Woodward's *Insdustrial Organization,* Perrow developed a two-dimensional typology which classifies organizations into four groups.[1] The two important characteristics recognized by Perrow are: 1) the degree to which the task performed is routinized and has few exceptions and 2) the degree to which the task performed is based on analyzable principles and known ways of problem solving as opposed to techniques which re-invent "new" solutions every time variations of a problem is posed.[2] Based on these two characteristics, Perrow recognizes four types of production technologies, each characterized by a particular type of organizational structure which he calls: craft (decentralized), non-routine manufacturing (flexible, polycentralized), engineering (flexible, centralized) and, routine continuous process

---

[1] Joan Woodward. *Industrial Organization* (London: Oxford University Press, 1965).

[2] C.B. Perrow. *Organizational Analysis: A Sociological View* (Belmont, CA: Brooks-Cole, 1970), Chapter 3.

(formal, centralized). Perrow concludes his chapter (Chapter 3) on *Bureaucracy, Structure, and Technology* by stating:

> However, the degree of bureaucratization does vary among organizations. Some organizations are willing to pay the price of high unit costs and to forego some of the economies of specialization and standardization. They do so because so little is known about designing, producing, and distributing the product that they have no choice; or because the environment is highly unstable; or because there is a demand for customized, high-quality goods. These organizations are less structured—or less bureaucratic—because they lack the characteristics that would make bureaucratization efficient.
>
> We argued that the best way to conceive of these two types of organizations was in terms of their technology, which may be either routine or nonroutine. If the technology can be made routine (because of sufficient knowledge about designing, production, and marketing goods and services and because there is large enough market to permit volume production), then a high degree of structure or bureaucratization is possible and is efficient. If the technology is not routine, then the organization must forego the advantages of high volume production and clear and elaborate structure; it will be less bureaucratic.[1]

As we shall see, very similar comments would apply to the degree of documentation required for the development of an ISO 9000 quality assurance system. Although I have only recently "discovered" Perrow's work, I mention it in passing because I find it interesting to note that his typology closely parallels my own observations derived over the past four years while helping various companies "implement ISO." Recognizing that there are several categories of companies may at first

---

[1] Perrow, op. cit., p. 90.

seem trivial, yet it is important because *it implies that there is not one implementation model.* This must be emphasized because all too often consulting firms, "consultants" AND some registrars simply rely on their most recent experience(s) and/or are influenced by past activities to develop and enforce THE(IR) paradigm for ISO implementation.[1]

Coming up with a detailed typology of American industry would require much research, time and effort. Such is not the writer's objective in a work of this nature. My purpose is to propose a low resolution (few details) typology. This model should help ISO "implementers" decide how to interpret each ISO clause.

ISO applicants can be broken down into three major categories: *Manufacturing, Distributing and Servicing.* Manufacturing firms include tens of thousands of companies which collectively either fabricate or assemble millions of products. Distributors collect manufactured goods, may modify or do partial assembly and distribute those goods to customers. Finally, the service industry includes just about everything else from the engineering consulting firm, which either designs processes, or plants or both, down to the professional service agency, which provides the part-time help sought by engineering consulting firm. Each cluster of industries in turn operates vastly different processes.

---

[1] This phenomenon can be observed time after time when an individual who has helped his/her company achieve "ISO registration" is immediately "pirated away" from his/her firm by a consulting firm. The newly hired "consultant" begins his career by lecturing to many audiences telling how "his" company achieved ISO registration. The problem with such an approach is that, although some common themes can be found across companies, the experience of one company does not necessarily translate well for other companies. For example, the traceability required for a supplier to the military is much more demanding than that required for a manufacturer of tennis shoes, yet the consultant will often state that the same traceability is required!

Types of Processes

Processes come in a variety of sizes, shapes, functions and purposes. There are continuous processes, batch processes, automated processes, semi-automated processes, manual processes, CAD/CAM processes, N.C. processes, federally regulated processes (FDA, DOT, FAA, FCC, etc.) and more. The work force which monitors these processes ranges from highly skilled full-time technicians operating a sophisticated computer system within the confines of a control room to part-time workers with no or little high school education performing routine, yet crucial, tasks.

The flow characteristic of a process—that is, whether it is a continuous or discrete process—can be a function of the time increment during which it is analyzed. A bottling assembly line processing tens of thousands of discrete units (cans, cartons or bottles for example) an hour, could be defined as a continuous process. Generally speaking, however, the distinction between a continuous and a discrete process is determined by the nature of the product being processed. Thus, in the case of a bottling plant, although the processing of the liquid filling the units came from a continuous process, the bottling process itself is "nearly" continuous. At the extreme end of the process spectrum, one finds manufacturers of very short runs (not in time but in units produced) such as, for example, ship yards or manufacturers of highly specialized (and expensive) one-of-a-kind prototypes, such as lasers or customized instruments.

Manufacturing processes can thus produce thousands of low-priced components a minute with low value added, such as plastic bottles or bottle caps, or a few highly customized 100 million dollars units a month with much value added (e.g., airplanes).

Within the service industry much the same distinction applies. Some "restaurant" chains specialize in selling hundreds of low-priced hamburgers an hour, whereas others specialize in offering a few dearly priced items per evening. Some service industries, such as suppliers of temporary help, act as matchmaker by simultaneously dealing with two customers. On the one hand they must satisfy the individual searching for a part-time job, and on the other hand they must please the company looking for the right part-timer.

The plethora of processes and human skills associated with each operation, of which we have only uncovered the tip of the iceberg, does lead to some important questions regarding the implementation of ISO. Indeed, do each of the ISO paragraphs have the same meaning to the manufacturer of whiskey (a continuous process), and the manufacturer of propylene, also a continuous process? Let us examine a few examples.

What Do You Mean by Contract Review?

The publication by the American Society for Quality Control of *ANSI/ASQC Q90 ISO 9000 Guidelines for Use by the Chemical and Process Industries* (1992), seem to indicate that potential users of the ISO standards do need guidance on how to interpret and bring relevancy to ISO's clauses. One cannot help but wonder if, or more likely, when, more guidelines will begin to appear. Will we begin to see guidelines for the bottling industry, meat packing industry, dry cleaning industry, food industry, software industry, pulp and whatever else industry? Perhaps, although I question whether there really is a need for the proliferation of highly specialized guidelines aimed at satisfying one minutely defined industry after another. Naturally, as long as there is a (micro)-market for a topic, why not satisfy that market. Indeed, this very book could be viewed as a guideline, but at least its intended audience is the community of ISO "implementers" at large.

When interpreting ISO clauses I believe the best policy is to first use common sense. Let us examine some examples. The *Contract Review* clause (4.3, ISO 9001) is a good starting point.[1] The clause basically states that the supplier shall review contracts to ensure that:

a) the requirements are adequately defined and documented;
b) any requirements differing from those in the tender are resolved;
c) the supplier has the capability to meet contractual requirements.

Naturally, as is the case with most ISO clauses, paragraph 4.16 (*Quality Records*) is invoked. In this case, the clause specifies that "records of such contract *reviews* shall be maintained."

When faced with clause 4.3, some individuals invariably ask "What do you mean by contract?" A contract as defined by Webster's *New World Dictionary* is "an agreement, especially a written one enforceable by law, between two or more people." Since the length, content and syntax of a contract will vary according to the product/customer demands, interpretation of paragraph 4.3 will naturally vary.

How would a restaurant wishing to apply for ISO registration interpret 4.3 (which ISO standard would apply, 9001, 9002 or 9003?). Would the interpretation be the same for a hotel manager, a dentist, a hospital director, or a software company? Obviously not, and yet there are similarities. In a restaurant, the contract review is performed at the customer/waiter(ess) interface. Customers order from a catalogue (better known as a menu). Requirements differing from the "tender" are resolved on the spot with the full confidence that the kitchen will be able to meet the modified contractual requirement (sales people are well known to have full confidence in the unlimited skills of their manufacturing facilities). In the rare event that the kitchen (where paragraph 4.9 *Process Control* would take effect) cannot deliver what was promised by the waiter(ess), a nonconforming meal (paragraphs

---

[1] All ISO references are taken from the ANSI/ASQC Q9001 Series.

4.13 *Control of Nonconforming Products* and 4.13.1 *Nonconformity Review and Disposition*) will be the end product. However, as is the case with some manufacturing industries, meal nonconformities can only be resolved *after* the product is delivered. Most restaurants will gladly apply corrective action (4.14) by either replacing or otherwise adjusting the order at no extra cost to the customer.

As for the requirement to keep quality records, records of all contract reviews are written on the ordering ticket which is probably retained (see 4.16) for no more than a few days at the most. Consequently, we see that service organizations can indeed bring relevance to most if not all of the twenty ISO 9001 paragraphs.

A hotel manager is likely to interpret 4.3 in a slightly different manner. The contract review process between the hotel's representative and the potential customer is likely to take place over the check-in counter. If the customer likes the price (which can often be negotiated, particularly during the off season), the room is "rented" for a predetermined period of time. The contract review process is not terminated, however, because the customer must also be satisfied with his/her room and the room service staff (clause 4.19 *Servicing*)—although that is usually implied and not specified in the "contract." To improve service, some hotels have empowered their staff, including room service staff, to directly handle some negotiations with customers. As for clauses 4.13 and 4.14 (*Control of Nonconforming Product* and *Corrective Action*), they are handled much as a restaurant would. In fact, most clauses can find a direct and practical application within the hotel industry. The exception or non-applicability might include clauses 4.8 *Product Identification and Traceability*, and 4.7 *Purchaser Supplied Product*.

The software manufacturer might have the greatest difficulty satisfying paragraph 4.3 because in many cases customer requirements are constantly changing as the software is being developed. In some cases this is simply due to the fact that the customer might not have a clear idea of what is needed until *after* the software is delivered. In other cases, the programmer tries to anticipate what the ill-defined requirements might be and then spends a considerable amount of time convincing the customer that the current program structure is really what the customer wanted in the first place. Thus, although software methodologies have been developed over the past ten to fifteen years to better design and develop software (i.e., CASE technology, for example), the need for contract review and the development of software metrics to monitor software performance/quality is particularly relevant in this industry.[1]

What Do You Mean by Nonconformities? (Review and Disposition)

All suppliers will have to address this particular clause (4.13) which calls for, among other things, the assessment of the nonconformance to determine whether it needs to be reworked, accepted with or without repair by concession, re-graded, rejected or scrapped (refer to standard). The following newspaper heading: "Contaminated fuel may have caused skydivers' crash", is a constant reminder of the importance of the proper testing and disposition of nonconforming material.[2] In this particular instance, the fuel became contaminated when stored in a truck which obviously had not been properly cleaned (see discussion on Assessment of Sub-Contractors).

For some suppliers, such as the manufacturer of whiskey, the "detection" of nonconforming product might well take several years. When the quality director learns about the problem, it might well be that the nonconformance had nothing to do with the original quality of the product. Whiskey is usually sold to distributors who store the product or, in some cases, even bottle it. If the distributor "mishandles" the bottling process, the product can slowly deteriorate

---

[1] See for example Edward Yourdon. *Decline and Fall of the American Programmer* (Englewood Cliffs: Yourdan Press, Prentice Hall, 1992), particularly Chapters 7 and 8.

[2] Sixteen people were killed during the crash at the Perris Valley Airport in Southern California.

soon after it is delivered. In such cases, the FDA requires the supplier to withdraw the product from the shelf.

As for quality records and retention, this same whiskey manufacturer had detailed quality records going back to 1909, obviously much longer than what would be required of a restaurant, for example.

Some companies would not even admit to making out of specification material. They simply re-grade or re-classify the product for alternative clients or applications. Such practices can lead to some rather impressive costs of "poor" quality. One company the author is particularly familiar with has come to rely so much on the re-classification of its products that it now subcontracts for the *storage* of over 14 million pounds, some of which is as much as three years old. A classic example of the so-called hidden factory, or rather the *hidden storage*, out of sight out of mind!

Finally, one may have to distinguish between nonconformities discovered internally and those discovered, by customers after shipping. As everyone knows, it usually is cheaper and less embarrassing to discover nonconformities prior to shipping. With regard to nonconformities, it is surprising how few companies keep *relevant* cost figures of "poor quality" or even bother to analyze these figures. I am not suggesting that U.S. accountants do not maintain such records, for indeed, over the years, I have seen many tables which breakdown the various labor and material costs associated with nonconformities. However, few of these tables seemed to ever be analyzed (Pareto Charts for example), or even verified for accuracy. As is well known there are many creative ways to account for nonconformances. Although the cost of "poor quality" is not specified in 9001, 9002 or 9003, it is mentioned in 9004 and will no doubt be included in future ISO revisions (besides, some of your customers might already require you to monitor your internal and warranty costs).

Design and Document Control (4.4)

4.4.1. General

> The supplier shall establish and maintain procedures to control and verify the design of the product in order to ensure that the specified requirements are met.

Design control only affects companies that wish to achieve ISO 9001 registration. The clause consists of eight sub-clauses subtitled: *General, Design and Development Planning, Activity Assignment, Organizational and Technical Interfaces, Design Input, Design Output, Design Verification and Design Changes.* It is a most important clause because one would hope that, in the words of Donald A. Norman, "Industry has learned that quality has to be designed into a product: insisting on proper design and manufacturing techniques from the very beginnings is far more effective than attempts to discover and repair ill-made goods on the production line."[1] The easiest and most simplistic way to interpret this clause is to require every ISO 9001 candidate to abide by every "shall" clause listed under paragraph 4.4. However there can be some surprises. The design process is, as most other processes, a multifunctional activity which involves several players (4.4.2.1 and 4.4.2.1). At a minimum, as many as four or more major parties are involved: the supplier's sales person, the customer, the designer(s), the manufacturer and, in some cases, technical services and one or more regulatory agencies (see 4.4.4c). In many cases, marketing is also actively involved in the process, interacting with both designers and the customer. Marketing's role and the nature of its relationship with other players will vary from company to company.

---

[1] Donald A. Norman, *The Design of Everyday Things* (New York: Doubleday/Currency Edition, 1991), p. viii. From an engineering point of view, see also John R. Dixon's *Design Engineering: Inventiveness, Analysis and Decision Making* (New York: McGraw-Hill Cook Company, 1966).

One must also distinguish between the types of designs as well as whether the design involves a new product or modification of an old product. Does your company design office furniture, buildings, home appliances, chemical products (such as ink additives or oxidizing agents), software, computer keyboards, books (contents) or book covers, hardware interfaces, precision bearings, questionnaires (yes, good questionnaires must first be designed and tested), etc? For some designs, functional characteristics of ergonomics are important (e.g., furniture), for others, ergonomics or esthetics are not as important (e.g., precision bearings). In the early design stages of a new product or even design modifications of a "catalogued" product, customer or marketing input characteristics/parameters can consist of vague specifications: "must be comfortable", "must not dry too quickly", "must have a high enough coefficient of friction", etc. Sooner or later (the sooner, the better), these fuzzy specs should be operationalized in terms of input and output parameters that can be calculated and measured (4.3 and 4.4).

When documenting the design process (second tier documentation), it is always a good idea to include a schematic of the design process. The process usually involves several departments and requires close collaboration and interaction with internal (technical services, marketing, sales, process engineer, quality control), as well as external customers.

Whereas some suppliers strictly design per customer requirements offering minor recommendations, other suppliers actively take part in the design process and may even own or subcontract some phases of the design including *design verification* (4.4.5). Some suppliers (casting companies, for example) may only design the manufacturing process and identify key monitoring parameters; however, manufacturing manufacturing specifications are strictly controlled by the customer. Other suppliers may have to design a machine around a product. The machine must be able to produce so many thousands of units per hour (diapers for example), under a broad range of environmental conditions using narrow raw material specifications. In such cases the machine usually ends up being a prototype which requires months of fine tuning involving *design changes* (4.4.6) until it is finally delivered thousands of miles away, perhaps never to be seen again. Contrary to what might be thought, design verification, records and change control are not always well maintained and supervised in such an environment—the argument being that "we'll never design another one like this again anyway!". One would think that much knowledge and process information is surely lost for ever, only to be rediscovered over and over again.

Nevertheless, the machine in question will likely require servicing (4.19, ISO 9001) which is usually subcontracted, particularly if the equipment is sold overseas. In such cases, paragraph 4.19 (*Servicing*) may become quite involved (*if specified in the contract*) and *might* include such activities as: technicians' training (4.18, *Training*), translation of service manual and document control of all such manuals (4.5.1 *Document Approval and Issue* and 4.5.2 *Document Changes/Modifications*) and maintenance records. Companies involved with the maintenance of equipment throughout the world—medical scanners or geological exploration sensors, to name but two examples, have attempted to solve this difficult problem by conducting all technical training sessions in English.

Multinational software companies have long had to translate software instructions (menus for example), for each foreign market. In such cases, document control and accuracy becomes a particularly relevant topic.

In another case, the design process might only take a few hours or a few days. A customer wants to test a new idea or (more likely) modify a process. To do so however he may need a new or slightly modified piece of equipment to *test* his idea. Not wanting to apply for all sorts of regulatory permits, he must find a supplier which can deliver the "new" equipment within a few days (usually five days). In such a case, although most of the design control activities are performed, they must be completed within a few hours.

For the software (or any other) industry, the degree of design control will vary according to the company's maturity. Yourdon, referring to the Software Engineering Institute model (SEI), identifies five levels of maturity: (1) initial, (2) repeatable, (3) defined, (4) managed and (5) optimizing.[1] Level 1 organizations have not formalized the design process and invariably rush the coding process and debug their programs on an "as found" basis. Level 2-4 organizations are at various stages of process repeatability. These companies usually have very detailed design review phases and some even have a separate department in charge of monitoring design phases. Processes are assigned to teams, and project leaders keep each other well informed as to the status of the "deliverables." As anyone who has used canned software will testify, this detail design process does not guarantee that the product will be error free. Hence the use of "Read Me" files to inform users of updates or work-arounds (software jargon for letting the user handle corrective actions). Level 5 firms have reached the optimizing level and emphasize ongoing continuous improvement via various feedback mechanisms.

The design phase will therefore be handled differently depending on the industry. Some will find that industry standards far exceed the ISO requirements. Others will discover that efforts will have to be devoted to develop a new set of procedures. Still others will argue that the clause cannot be applied to their industry (without significantly reducing profit).

Document Control (4.5.1 and 4.5.2)

One of the most difficult paragraphs to comply with (at least for a great many companies), deals with document approval, issue, changes and modifications. The clause states that all documents relating to ISO requirements shall be controlled to ensure, among other things, that

---

1 Edward Yourdon. *The Decline and Fall of the American Programmer.* Englewood Cliffs: Yourdon Press, Prentice Hall Building, 1992 Chapter 4, pp. 74-83.

pertinent information is available at all locations "where operations essential to the effective functioning of the quality system are performed." Moreover, the clause specifies that obsolete documents shall be "promptly removed from all points of issue or use." Any modification shall be reviewed, identified and approved.

The need to control documents, particularly drawings, was recognized as long ago as 1908 by Ford. The following quotation from David A. Houndshell's book succinctly summarizes the importance and integral part of document control within a quality assurance system:

> The company produced and maintained drawings of every part of the Model T, every special tool, jig, fixture, and gauge used in its production, and every master gauge used to check these special devices. Drawings served as the ultimate authority in Ford production, for they specified dimensions, tolerances, gauging points, materials (including shear strength and other metallurgical specifications), and finishes. Used by the design, tool, engineering, and inspection departments, superintendents . . . and foremen in each of the parts production departments, these drawings served as the medium for exchanging information and for maintaining common understanding. *No changes could be made without a change in the drawing.* . . . Change acted like a pebble hitting the middle of a still pond; ripples moved out to the various departments mentioned above.[1]

Houndshell's description of the need for document control as it applied to Ford's Model T is still relevant today. However, although the need for document control can hardly be questioned, the degree

---

[1] David A. Hounshell. *From the American System to Mass Production 1800-1932* (Baltimore: The John Hopkins University Press, 1984), pp. 272-273, emphasis added.

of emphasis will vary from organization to organization. Thus, whereas it is true that enterprises operating within the pharmaceutical, medical, food, avionics, aerospace, nuclear and related industries are required to maintain elaborate document control procedures, other less-regulated industries will not likely be required to maintain similar systems.

The requirement to ensure that obsolete documents be "promptly removed from all points of issue or use," is not an easy task to achieve. It is indeed difficult to prevent people from photocopying documents. Some ingenious ideas include: printing all controlled documents on special company paper (or colored paper) or, transferring all documents to diskettes or to a centralized mainframe. Both approaches have their advantages and minor disadvantages. Printing on company or colored paper reduces but does not really eliminate the risk of photocopies. The use of "read-only" diskettes does attempt to control documents. Some customers issue their quality manuals and everything else on "read-only" diskettes. The transfer of documents to a mainframe perhaps comes closest to attempting a foolproof document control.

When transporting a document system to a centralized mainframe, a couple of security procedures will have to be addressed:

- Password hierarchy (access code). The decision will have to be made as to who can access and modify what document(s) (operators, process engineers only, managers, etc.).

- How to handle/control printed documents. Some companies print a disclaimer on all pages. The disclaimer, usually a header or footer, states that the document is only valid for the current date.

- Inter-departmental access. How will departments access each others' documents?

Process Control (4.9)

The two clauses (4.9.1 General and 4.9.2 Special Processes) which comprise paragraph 4.9 are not particularly long, yet they may well require a very significant amount of effort. To ensure that processes are operating under control conditions, paragraph 4.9.1 calls for:

- Documented work instructions defining the manner of production and installation, *where the absence of such instruction would adversely affect quality*

- Monitoring and control of suitable process and product characteristics during production and installation

- Approval of processes and equipment, as appropriate

- Criteria of workmanship to be stipulated, as need be, in written standards or via representative samples

The need for documented procedures cannot be overemphasized, particularly when such procedures may well affect a patient's life. A press release dated August 19, 1991 began with the following words: "An unidentified heart patient had his third heart in four days Sunday." Steve Marshall, author of the article, explains that "[T]he man who has Type O blood, received a heart Wednesday from a donor with Type A . . . The hospital's procedures met national standards, but more checks have been implemented 'to make sure this never happens again. Specifically, we will now require additional blood-type testing and a second separate communication between the donating hospital and organ retrieval team'." [1] Clearly, in some cases, national standards are not sufficient.

The need for operating instructions or operating sheets has long been recognized by manufacturers. During the 1880s, the Singer company implemented at its Elizabethport factory the use of a "blue book." The book, written by Singer's production experts, "delineated all of the machining operations and work-flow routes for the Improved Family sewing machine. [It] also specified inspection procedures and limits of precision. In sum, it codified Singer factory production operations for the first time in the company's history, and it made clear—implicitly, at least—*that control over these operations had been assumed by Singer factory managers*."[2]

If the above procedures, implemented over 110 years ago at Singer, were indeed followed (as they seem to have been), Singer's blue book system is likely to have impressed any 19th or 20th century registrar. Of particular significance to today's application is the reference by Hounshell of Singer's management having assumed control of operations. One might be inclined to believe that Hounshell meant to suggest that Singer's management *regained* control of the means of production from the factory workers. Whether that is good or bad is difficult to assess, for it all depends on your point of view. Singer's experience was certainly positive, for it was one of the many first endeavors designed to achieve that most elusive goal of part interchangeability and thus repeatability. Certainly, one of the crucial steps required to achieve product or service repeatability is the standardization of work instructions and procedures. Whether management's control of processes is the only or even desirable means of achieving such process control is debatable.

---

[1] Steve Marshall, "Transplant corrects heart mix-up," USA TODAY. August 19, 1991.
[2] Quoted in David A. Hounshell. *From the American System to Mass Production 1800-1932* (Baltimore: The John Hopkins University Press), 1984, p. 120, emphasis added. Operation sheets were also used by the Ford Company as early as 1908, Hounshell, op. cit., p. 273.

What we learn from Houndshell's account is that, in the early years of industrialization (as is still true today), whatever processes control might have been exercised, it was likely to be either modified by skilled but little trained operators, continuously improved to reach the ever elusive optimum operating characteristics or, more likely, a combination of both. In that sense, process control was and continues to be in the hands of those that *work* with it on a daily basis. Based on that premise, it is therefore important to focus on which work instructions need to be documented, how and by whom.

How and What to Document, and Who Shall Do the Writing?

What work instructions need to be documented can only be decided by individuals working at the site. When I am asked by clients: "What do you think we need to document to satisfy ISO?", my answer is invariably: "It's not a question of documenting for ISO but rather documenting to ensure, within reason, a standardized form of quality." What do I mean by "within reason." First of all, there is no need to believe that you will have to document *everything*. In the early days of ISO 9000 consultation, consultants and some registrars would insist that ALL processes and operations be documented. That is not what the standard states. One of the key sentences in paragraph 4.9.1 states: "where the absence of such instructions would adversely affect quality." Nowhere does it imply or specify that you must document everything. And yet, that is precisely what some companies chose to do. Not realizing that documentation can be an expensive proposition, some companies invest thousands of man-hours to document every single process. One 1,200 employee company I know of, estimates that it has invested as much as 2,500 man-hours or more to implement an ISO 9002 quality assurance program. Everything, including the company's breath analyzer calibration procedure has been documented. Although it is true that insurance companies require such equipment to be calibrated, it is difficult to rationalize the calibration of a breath analyzer within the scope of an ISO 9002 audit.[1] Nonetheless, the company in question felt it benefited greatly from the exercise. That is all the justification that is required.

When addressing the topic of "documented work instructions" and the "monitoring of process and product characteristics," the best strategy to adopt is to include the work force in the documentation efforts. Alienating the work force is a sure way to invite nonconformities. When reviewing a process, simply ask operators how they do it and why they do it that way. This will allow you to review deviations from standard operating procedures, *discuss* (with operators) and *modify* and thus *update* said S.O.Ps; as well as determine what educational and/or technical training might be required to ensure that the procedures are *understood* and thus *accepted* and *followed*.

Procedures are usually not documented simply because the operators have so many years of "experience." Yet, when operators are interviewed, they often will confess of not knowing why things are done a particular way. No doubt, they all have experience and know what needs to be done when, but they often do not fully understand why it is important that a particular set of instruction is done in a particular order. The impact to their internal customers downstream is not always appreciated. The reason why this is so is simple: nobody has really explained to them why it is important (see Training (4.18), below). Of course, all operators receive on the job training, but that so-called training, accumulated after years of *observations*, generally consists of a set of very specific instructions designed to prepare the operator to react to a particular set of conditions which can, in some cases, number in the hundreds. The "big picture" is thus never really presented to most operators except of course to those that have worked upstream and/or downstream from their operation.

Processes are thus monitored by two types of individuals: process engineers who have a holistic (but not necessarily deterministic)

---

1 I suppose the breath analyzer might be used to determine whether or not an employee's performance might affect product quality.

knowledge of *why* and *how* a process operates and operators who have a compartmentalized conception of how a process might behave at a particular time and place. The one individual has (hopefully) a scientific knowledge of a whole set of processes; the other has a pseudo-scientific comprehension of but one or two sub-processes. This continuous dichotomy between those that know *why* and *how* something functions and those that know *how* something behaves needs to be somehow married into one operational document. If a process is documented by the process engineer, it is likely to be too complicated for the operator. If the process is documented by an operator it is likely to skip a few fundamental steps. The compromise, a generic process flow diagram (or process sheet), attempts to meet the dual objective of satisfying the process engineer and the operator simultaneously.

The Generic Process Flow Diagram

When an organization attempts to flow chart its processes, it quickly finds out that hundreds of processes need to be documented. In many cases, the processes are "children" of parent processes each monitored by tens of parameters. An attempt to document each process and maintain document control procedures will quickly discourage anyone. One option is to use generic flow charts.

The methodology of generic flow charts or flow diagrams is derived from data flow diagrams first developed by Tom De Marco. The hierarchical methodology summarized in Figure 3.0 below begins from a top level *context diagram* which consists of several first level nodes. Each first level node is in turn expanded into as many levels as required (see also Figures 3.1 and 3.2).[1]

---

[1] Tom De Marco. *Structured Analysis and System Specification* (Englewood Cliffs: Yourdon Press, Prentice Hall Building, 1979). See also Alan S. Fisher. *Case: Using Software Development Tools* (New York: John Wiley & Sons, Inc., 1991).

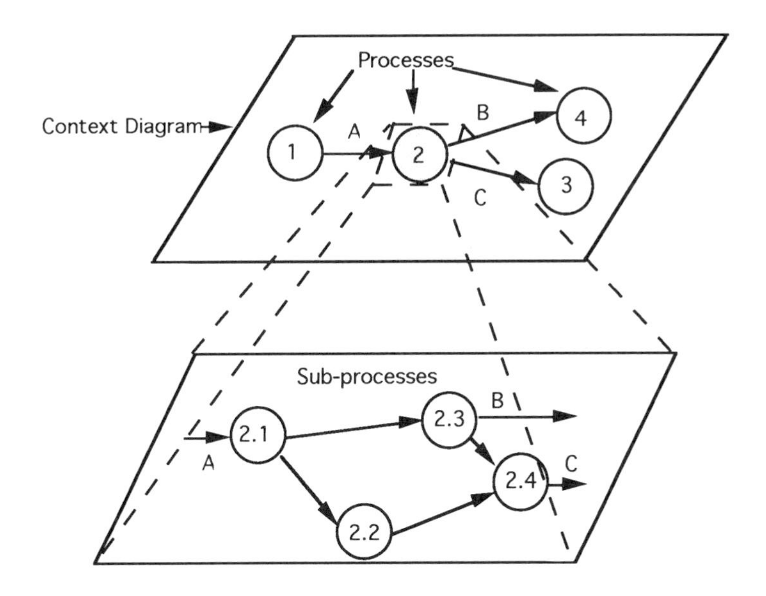

Figure 3.0 Data Flow Diagram

Assuming that the number and type of processes that need to be documented has been established (not an easy task), the next step is to categorize these processes into three to seven major categories. For each major category, outline a general schematic of the process identifying all key parameters. *Be careful to note that the identification of these key process and product parameters implies the need to monitor and thus control said parameters. Moreover, this will also imply the need to monitor (calibrate, maintain, etc.) instruments assigned to the particular measurement.* One way to quickly identify key product characteristics is to check customer specifications. These customer (product) specifications will in turn generate a whole set of key process/product specifications usually identified by various sampling points spread throughout the process.

By identifying all key parameters, you simultaneously identify all key process instruments (e.g., on-line analyzers, thermocouples, voltmeters). Having completed the task you should have identified a *subset* of all your measuring points (I am excluding laboratory analysis from this discussion, see chapter 3). The size of this subset will vary from industry to industry (half a dozen parameters/instruments for some, two to three dozens for others). Figures 3.1 and 3.2 illustrates the generic flow chart process. Figure 3.1 is a schematic representation of a process which consists of six sub-processes. Figure 3.2 represents a telescoping view of sub-process 1.0. This telescoping approach to flowcharting can be iterated as needed. Notice that no specifications as to temperature and flow are mentioned on the flowchart. Such specifications can usually be obtained from work sheets issued daily or weekly. The purpose of the generic flow chart is to provide a schematic, which incidentally could be used for training, of how the process flows, what needs to be monitored and what corrective action(s) needs to be taken. Simple as it is, the generic flowchart can prove to be very helpful when presenting the "big picture" to experienced operators.

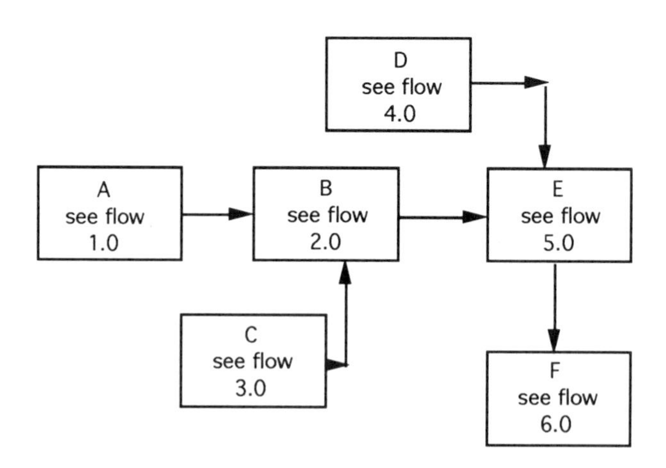

Figure 3.1 Generic Block Diagram of a Simple Process

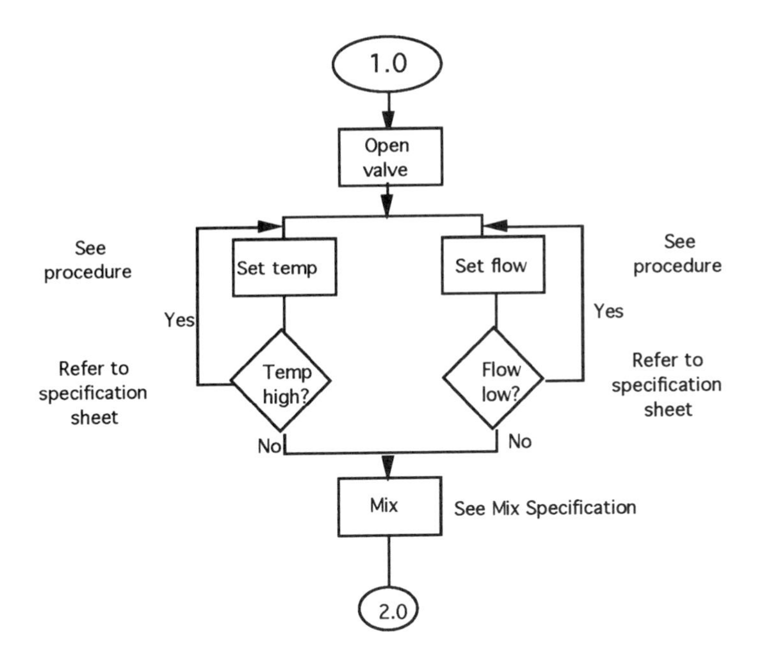

Figure 3.2 Sub-process 1.0

Inspection, Measuring, and Test Equipment (4.11)

> *Measurement: The quantitative determination of a physical magnitude by comparison with a fixed magnitude adopted as the standard, or by means of a* ***calibrated*** *instrument. The result of measurement is thus a numerical value expressing the ratio between the magnitude under examination and a* ***standard*** *magnitude regarded as a unit.*[1]

The above definition of measurement which emphasizes calibration and reference to a standard, summarizes some of the intent of paragraph 4.11 (ISO 9001). Clause 4.11 is the longest of all ISO clauses and can be for some organizations (except perhaps those

[1] From *The New Caxton Encyclopedia* (1969), quoted in P.H. Sydenham, *Measuring Instruments: Tools of Knowledge and Control* (London: Peter Peregrinus Ltd., p. 10), emphasis added.

operating under MIL-STD-45662A or similar standards), one of the most difficult to implement. This is partly due to the fact that little guidance is provided within the standard and within the ISO 9004 guidelines as to the importance and cost of measurement as well as the necessity of establishing an instrument calibration and precision assessment program. However, one cannot really blame the standard, the fact remains that for too many individuals, clause 4.11 is either not relevant or much too expensive to enforce. As far as these individuals are concerned, once purchased and installed, instruments are to be left alone. Yet, there is much more to it.

The need to measure in order to acquire information and thus knowledge about a process was perhaps first expressed by Lord Kelvin when he stated:

> In physical science a first essential step in the direction of learning any subject is to find principles of numerical reckoning and methods for practicably measuring some quality connected with it. I often say that when you can measure what you are speaking about, and express it in numbers, you know something about it; but when you cannot measure it, when you cannot express it in numbers, your knowledge is of a meager and unsatisfactory kind: it may be the beginning of knowledge, but you have scarcely, in your thoughts, advanced to the stage of science, whatever the matter may be.[1]

---

[1] Lord Kelvin's speech to the Institution of Civil Engineers on May 3, 1883, cited in P.H. Sydenham, *Measuring Instruments: Tools of Knowledge and Control* (London: Peter Peregrinus Ltd., 1979), p. 4. Sydenham also cited V. Papanek who states: "perhaps the oldest of man's intellectual myths is the one that links measurement with truth and justice. A modern manifestation of the myth is in the craze for quantization. Let's pin numbers to everything because that makes grading easier!" Papanek, *Design for the Real World: Making to Measure* (London: Thames & Hudson, 1975). Quoted in Sydenham, op. cit., p. 5.

Over the last one hundred or so years since Lord Kelvin's speech, the need to measure has certainly been recognized by most. However, it is debatable whether or not individuals involved with monitoring or measuring processes or product characteristics understand the cost association with any measurement process. Realizing that the cost of a wrong measurement may lead to disastrous consequences (lawsuits) for a company, Sydenham stresses the importance of establishing good measurement and identifies a dozen steps directly or indirectly associated with the costs of measurement. These are:

1. Determining that there is a need to measure
2. Deciding what to measure
3. Evaluating the type of instrument required
4. Conducting some experiments (optional)
5. Choosing which instrument and from which vendor
6. Placing the order
7. Calibrating the instrument (accreditation may be required)
8. Familiarization with the instrument
9. Installation (where along the process?)
10. Updating plant drawings
11. Writing operators' instructions
12. Establishing service, calibration and maintaenance programs to help assess Mean Time Between Failure[1]

All of the above suggestions directly relate to the intent of clause 4.11 which calls for—among other things—for the supplier to:

- Identify the measurements to be made, the accuracy required, and select the appropriate inspection, measuring, and test equipment

- Identify, calibrate, and adjust all inspection, measuring and test equipment, and devices that can affect product quality at prescribed intervals

---

1 Sydenham, op. cit., pp, 131-141.

• Establish, document and maintain calibration procedures

• Ensure that the inspection, measuring, and test equipment is capable of the accuracy and precision necessary

• Identify inspection, measuring, and test equipment with a suitable indicator *or approved identification record to show the calibration status*; [Note: the italicized wording seems to be ignored by people who go through the futile exercise of trying to attach a label to every instrument (some use ingenious color schemes such as red month, yellow month, etc.). This approach may well be practical within a laboratory setting but it often becomes impractical to physically label thousands of greasy gauges with stickers which do not stick! I do realize that such requirements are often imposed by the customer.]

• Maintain calibration records for inspection, measuring and test equipment[1]

• Other requirements relating to "suitable" environmental conditions, proper handling and other safeguard inspections (see standard for exact wording)

For most applications, the clause is intended to apply to gauges and instruments such as scales, calipers, pressure gauges, flow meters, thermometers, gas chromatographs, etc. There are nonetheless industries where inspection and measurement can only be achieved by ranking using an ordinal scale such as the Rockwell hardness test

---

1 Quoted from ANSI/ASQC Q91-1987, p. 5, paragraph 4.11.

or multiple comparison techniques favored by market research analysts. In such cases where measurements can only be based on a subjective preference based on the senses, the practical application of requirements listed in paragraph 4.11 (a-j) can be challenging. I was once asked by a seminar participant how paragraph 4.11 would apply in his industry where olfactory senses and taste buds were two critical "instruments." The industry in question was the distillery of fine spirits. After some initial discussions as to whether or not the clause was even relevant, we came to the conclusion that the principal taster (probably the one with the most experience and steadiest hand) would, in all likelihood, be considered the "standard nose" to which other tasters had to align/calibrate themselves with. I do recall that in the food industry, as well as other industries where taste or fragrance is of critical importance (the perfume industry, for example), statistical techniques have been designed over the years to standardize and differentiate/classify subjective parameters such as sweetness, texture, scent, etc. To my knowledge, the calibration of senses has not yet been addressed (experts might inform me otherwise). The precision and accuracy of olfactory, tactile and other senses has no doubt been studied and might even comprise a well-organized body of knowledge.[1]

It has been the author's experience that in most cases, individuals reading clause 4.11 will focus on the first few subparagraphs which emphasize the need to periodically calibrate instruments, maintain records of such calibration and be able to trace the calibration to some primary or reference standard.[2] Yet, assuming that these activities are

---

[1] Related works would include *Panel Surveys*, edtied by Daniel Kasprzyk, Greg Duncan, Graham Kalton and M.P. Singh (New York: John Wiley & Sons, 1989) and *Analysis of Complex Surveys*, edtied by C.J. Skinner, D. Holt and T.M.F. Smith (New York: John Wiley & Sons, 1989).

[2] Calibration is the process of comparing one standard against a higher-order standard of greater accuracy. The ability to relate measurements back to the primary reference standard is called traceability of the standard. For a brief overview of issues relating to Clause 4.11 (ISO 9001) refer to *Quality and Planning and Analysis: From Product Developement Through Use*, edited by J.M. Juran and Frank M. Gryna, Jr. (New York: McGraw-Hill, 1980), Chapter Sixteen, “Measurement,” pp. 386-406.

performed, which is not always true, clause 4.11 consists of much more than just calibration. One must also ensure that the instrument(s) are "capable of the accuracy and precision necessary" (4.11 d), as well as a host of other requirements (refer to 4.11). The apparently simple task of verifying that an instrument is capable of the accuracy and precision necessary can, in fact, be a difficult if not impossible task.[1] Indeed, it seems that, according to suppliers, some customers are wont to insist on specifications that far exceed the precision of any known equipment. A supplier of precision bearings once told me that his customer insisted on specs of plus or minus 25 millionth. Yet, although the supplier did agree to supply the parts, he was well aware that his instruments were not capable of such precision. The metrology department had performed some instrument capability studies and knew very well that given the above tolerance, the instrument's precision would nearly consume (over 85 percent) the whole specification interval! But why then agree to "unacceptable" specifications? "If we don't, someone else will, even though they too can't satisfy the requirements," was the answer. Similar stories can be told of laboratories which have to detect concentrations of less than 5 ppb (parts per billion), apparently well beyond the capability of some instruments.

A formal implementation of paragraph 4.11, including instrument capability studies such as gauge repeatability and reproducibility analyses, *might* help suppliers better inform their customers.[2] Customers may of course choose to ignore the information and still demand tolerances which will push instruments beyond their designed capability.

---

[1] See for example the *American Society for Testing and Materials (ASTM) Standards on Precision and Accuracy for Various Applications*, Philadelphia, 1977 or more recent editions.

[2] Gauge R & R studies can routinely be performed using a vast array of softwares. These softwares are advertised in the March issue of *Quality Progress*. An excellent reference is the Automotive Industry Action Group's (AIAG) *Measurement Systems Analysis. Reference Manual.* October 1990. 26200 Lahser Road, Suite 200, Southfield, MI 48034 Ph: (313) 358-3570.

Research or consulting organizations or marketing departments involved with the design and analysis of surveys will naturally interpret paragraph 4.11 very differently from a machine shop where jigs and fixtures are routinely used. In such cases, the instrument in question consists of a questionnaire whose reliability and discriminatory power must first be ascertained. Those versed in social science statistics will know that questionnaires can/should be tested on pilot audiences before being administered on a large scale.

As was the case with other clauses, the emphasis placed on paragraph 4.11 will depend on the organization's type of business. Robert Pennella, in his excellent *Managing the Metrology System*, states that "[T]he type of calibration system that a supplier must put into effect is dependent, of course, upon the complexity of the product design, contract quality requirements, and customer expectations."[1] A testing laboratory wishing to achieve ISO 9003 accreditation would have to rigorously implement each and everyone of the sub-paragraphs (a-j). In fact, the testing laboratory would have to also follow ISO/IEC Guide 25 *General requirements for the technical competence of of testing laboratories*, as well as ISO/IEC Guide 43 *Development and operation of laboratory proficiency testing* and ISO/DIS 10012 *Quality assurance requirements for measuring equipment*, a verbose *draft* document consisting mostly of definitions and more requirements. It is debatable whether a manufacturer of candies or ball point pens will need to implement as rigorous a program as a testing laboratory. This point may well be argued, however. Indeed, it has been suggested by some that the rigor with which paragraph 4.11 should be implemented is industry independent. Naturally, I do not share this myopic and impractical point of view, nor I hope will registrars.

---

[1] C. Robert Pennella, *Managing the Metrology System* (Milwaukee: ASQC Press, 1992), p. 5. Pennella's book is a concise book of some 85 pages in which the author reviews calibration system description, metrology audits and includes two case studies.

Quality Records (4.16)

Clause 4.16 (Quality Records), requires the supplier to establish and maintain legible, identifiable and retrievable records to "demonstrate achievement of the required quality and the effective operation of the quality system." The retention of such records, usually specified by state or federal regulations or by legal requirements, is to be specified. The clause, which specifies other minor requirements, would not be particularly difficult to enforce were it not for the fact that it is mentioned in thirteen of the twenty ISO 9001 paragraphs or sub-paragraphs. The requirements dictated by paragraph 4.16 will vary from industry to industry.

In some service industries, such as hospitals, careful maintenance of records is an important and stringent requirement which, if violated, may lead to dire consequences. In an article written by Martin Gottlieb for the *New York Times*, the author explains that "the organization that accredits a vast majority of the nation's hospitals voted today to prohibit any rewriting of important hospital quality-control records. The action came in the wake of accusations that some institutions altered the records to pass inspections and quality for millions of dollars in Federal Medicare and Medicaid reimbursements." The author explains that hospitals must keep minutes of highly confidential meetings during which doctors, nurses and other health-care professionals review their work. Gottlieb concludes by stating that "[H]ospitals caught changing the minutes risk loss of their accreditation, suspension from the accreditation process for one year, and referral of their cases to law-enforcement authorities."[1]

For other industries, failure to correct errors may lead to lawsuits. TRW Inc. recently had to settle a lawsuit with nineteen states and the Federal Trade Commission. Michael Miller of the *Wall Street Journal*

---

[1] Martin Gottlieb, "Hospitals Are Warned Not to Rewrite Records," *New York Times*, Sunday, May 10, 1992.

observed that "[T]he settlement represents a broad concession by TRW to widespread complaints that it routinely fills its credit reports with errors and ignores consumers' efforts to correct them." Quoting TRW's vice president for public affairs, the article continues by stating, "[A]s we listened to more people, we've become aware that at least to the consumer, following the letter of the law is not enough."[1]

Internal Quality Audits (4.17)

See Chapter 6.

Training (4.18)

Clause 4.18 (ISO 9001) requires the supplier to "establish and maintain procedures for identifying the training needs and provide for the training of all personnel performing activities affecting quality." In addition, persons "performing specific assigned tasks shall be qualified on the basis of appropriate education, training, and/or experience, as required." The clause concludes by stating that "appropriate records of training shall be maintained."

The lack of worker training programs in the U.S. has long been established. A recent investigation by the National Transportation Safety Board as to the cause of an airplane crash revealed that not only was contamination found in a truck from which the plane had been fueled but also that the pilot and co-pilot, both of whom had died in the crash, had no formal training in flying the De Havilland DHC-6 Twin Otter.[2]

---

[1] Michael W. Miller, "TRW Agrees to Overhaul Its Credit-Reporting Business," in the *Wall Street Journal*, Wednesday, December 11, 1991.

[2] One would think that a more rigorous *Receiving Inspection and Testing* (4.10.1) procedure might have to be implemented. A third-party auditor might want to look more carefully at the company's *Assessment of Sub-contractor* (4.6.1). This particular accident would seem to indicate that the truck company supplying the airport with fuel is not an acceptable sub-contractor.

This failure on the part of many American enterprises to emphasize and invest in employee training programs has been documented numerous times. A study by the U.S. Office of Technology Assessment found that whereas Japanese firms trained their employees more than 300 hours during the first six months, U.S. workers received less than 50 hours! The report also found little government encouragement for training and retraining. The average state training program assists just 64 companies and fewer than 4,000 workers annually. Most of the assistance goes to companies with more than 200 employees. Small firms, which account for 35 percent of U.S. jobs, often lack the funds necessary to train workers, whereas larger firms hire people with the required skills rather than train them.

The statistics quoted by the U.S. Office of Technology Assessment do *seem* to parallel my experience. I emphasize *seem* because with respect to training, as with most other issues relating to ISO 9000, the answer will vary depending on whom you talk to. Thus when management claims that training is provided, employees will counter by saying "what training?" It all depends on what is meant by training. Some companies have implemented intensive in-house technical training including improving arithmetic and reading skills. Others subcontract training with a local university or technical college. Within these companies, most white collar workers would unanimously agree that although there might be some technical training, including for example safety (OSHA related) training, there is very little if any "professional" career enhancement training. This claim would validate the above findings. Irrespective of the training status, one thing is certain, too many firms fail to maintain proper employee training records.

In one of the very few books devoted to the subject of on-the-job training, Martin M. Broadwell explains "[W]e should pay close attention to the *recording of the training* . . . Training records are a very valuable part of the total picture of a worker, and if we fail to keep these records up to date, we are presenting only part of the person to those who look at the records."[1] In a paragraph reminiscent of clause 4.18 of ISO 9001, Broadwell concludes by emphasizing that "they (records) should provide space to not only show what training has been done but *what training is anticipated.* Ideally, these records should show the step-by-step training program for each employee."[2]

Whenever training—orientation, on-the-job, off-the-job or outside training—is to be addressed by a company, it should recognize that an organization should commit its resources to a training activity "only if, in the best judgement of the managers, the training can be expected to *achieve some results other than modifying employee behavior.* It must support some *organizational end goal,* such as more efficient production or distribution of goods and services, reduction of operating costs, improved quality, or more effective personal relations."[3] One could of course add to the above quotation ISO 9000 implementation and training. The management of training issues can either be centralized within the human resources department or decentralized by department. At a minimum, what would be required would include:

- A list of employees, by department, specifying the type of training required (safety, certification, quality, general education, career enhancement, etc.).

- Each record *should* identify the course title, its type (video, lecture, seminar), its duration, the student's grade (optional) and when the course was

---

[1] Martin M. Broadwell, *The Supervisor and On-The-Job Training* (New York: Addison-Wesley, 1991), p. 149.

[2] Broadwell, op. cit, p. 150.

[3] Quoted from Ernest J. McCormick and Daniel R. Ilgen, *Industrial Psychology* (Englewood Cliffs, NJ: Prentice-Hall, Inc., 1980), p. 234. The authors are referring to the work of W. McGehee, "Traning and Development Theory, Policies, and Practices." In D. Yoder and H.G. Honenman, Jr. (Eds.), *ASPA Handbook of Personnel and Industrial Relations* (Chapter 5.1). (Washington D.C.: Bureau of National Affairs, 1970).

taken. Some companies even keep copies of the students "exams."[1]

- Additional information could include the validation period (six months, a year, etc.).

- Copies of certificate of completion, diploma, etc.

Conclusions

The purpose of this chapter has been to review and interpret some of the most difficult clauses. Naturally, the standards (9001, 9002 and 9003) contain additional clauses which have not been discussed in this chapter. These clauses are briefly reviewed in Appendix A. One important clause not yet reviewed is clause 4.1 *Management Responsibility*. Among the several subclauses listed in paragraph 4.1 is clause 4.1.1 *Quality Policy*, which states that the supplier shall ensure that the quality policy is "understood, implemented, and maintained at all levels in the organization." This often ignored clause is most important, for it not only states that a quality policy shall be stated (easy enough to do), but also that it shall be applied and practiced at all levels within the organization. For many companies this means having to do things differently; this in turn invariably means having to introduce changes, the very topic of our next chapter.

---

[1] Some regulatory agencies, notably OSHA, require records of exams for hazardous materials training.

4 How to Implement Change

"Three things make people want to change. One is that they hurt sufficiently . . . Another thing that makes people want to change is a slow type of despair called ennui, or boredom . . . A third thing that makes people want to change is the sudden discovery that they can." Thomas A. Harris, *I'm OK You're OK*, p. 60.

On Change

Statistics provided by registrars show that it takes on average six to fourteen months to achieve registration. The frequency distribution characterizing the time for registration is far from being symmetric; it is in fact skewed to the right. Thus, whereas very few companies manage to achieve registration within six months, most require anywhere from nine to fourteen months to get ready. The great variability in time is caused by many factors among which one can cite: size of the company, willingness to achieve registration, current state of readiness (i.e., current effectiveness of the quality assurance system), current state of flux (i.e. are new systems being installed, is the company about to be acquired, etc.), ability to organize and distribute assignments, amount of documents that will have to be generated or are believed to be necessary, ability to conduct effective internal audits, desire to achieve zero nonconformances prior to third party audit and, last but not least, the registrar itself. Although all of the above factors do contribute to a myriad of delays, none can be more devastating than the cumulative effects caused by people's innate ability to resist change.

The *American Heritage Dictionary* (1991) defines change as follows:

change b. To give a completely different form or appearance to; transform. **Synonyms:** *change, alter, vary, modify, transform, convert, transmute.* These verbs mean to make or become different. *Change* implies a fundamental difference or a substitution of one thing for another.[1]

For some companies, the implementation of any of the three ISO standards will require some form of *transformation* which might, in some cases, imply the need to *alter* or otherwise *modify* current business practices. Failure to recognize or communicate the fundamental need to operate differently can only lead to frustrations and unnecessary delays, not to mention an increased likelihood of failing the audit.

Types of Changes: First- and Second-Order Change

Among the many psychological and managerial/organizational books available on the subject of change, Paul Watzlawick's *Change: Principles of Problem Formation and Problem Resolution*, provides, in this writer's opinion, valuable observations.[2] Influenced by principles of cybernetics Logical Types and Group Theory, Watzlawick distinguishes between two types of changes: **first-order change**, that is *change that can occur within a system that itself stays invariant*, and **second-order change**, or *change of change;* the type of change which brings about a logical jump from one level to the another (see Figures 4.1 a and 4.2 b).

First-order change(s) brings about pseudo-transformations often referred to as the "more it changes the more it remains the same" symptom. Most of us have experienced such pseudo-transformations where a negative feedback allows a system to regain and maintain its internal stability. First order changes, or the maintenance of the status quo, are often experienced during ISO implementation. Examples of negative feedback mechanisms aimed at inducing first-order change would include "I don't have the time," "I don't know

---

[1] The *American Heritage Dictionary* (Boston: Houghton Mifflin Company, 1991).
[2] Paul Watzlawick, John Weakland and Richard Fisch, *Change: Principles of Problem Formation and Problem Resolution* (New York: Norton, 1974).

what you mean," "You don't understand how complicated it is," or its corollary, "It will take at least two years to implement." All of these arguments are usually put forth to counter a desire to bring about change. Thus to the request (action), "How long would it take to document our third tier?", the negative feedback reply (reaction), "Two man-years!" is stated in the hope that the request would "go away." Other examples of first-order change include the re-organization of yet another committee which will address, once again, ISO implementation. To induce genuine change, one needs to focus on a higher order of change, namely, second-order change.

Second-Order Change

Watzlawick points out that "the most pragmatic approach is not the question *why?* but *what?*; that is, what is being done here and how that serves to perpetuate the problem, and what can be done here and now to effect a change?"[1] For example, whenever a company experiences resistance to change, one should observe and note what are the reasons given to justify inactivity; in other words, one must look for the managerial antecedents which might have *caused* the (un)desired behavior to occur. All of the reasons (excuses) listed above have one common theme: *not enough time*. People are always willing to help, but, before squeezing yet another hour out of their busy schedule, they would like to know what compensation management is willing to offer.[2] There lies the snag. In order to bring about a *change of change* (second-order change), management must recognize that they too will have to change. Rather than expect something for nothing, concessions will have to be made. Delegating without participating is not sufficient. Indeed, if a company is genuinely interested in

---

[1] Watzlawick, op. cit., p. 86. See also Rosabeth Moss Kanter, *The Change Masters: Innovation and Entrepreneurship in the American Corporation* (New York: Simon & Schuster Touchstone Book, 1988).

[2] See Juliet B. Schor's *The Overworked American: The Unexpected Decline of Leisure* (New York, Basic Books, 1992).

achieving ISO registration, it must recognize that realistic deadlines must be set AND time must be allocated to achieve these deadlines.[1]

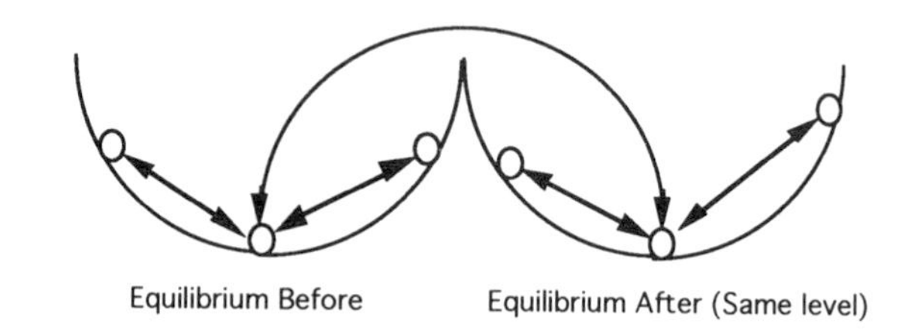

a. First-order change (more of the same)

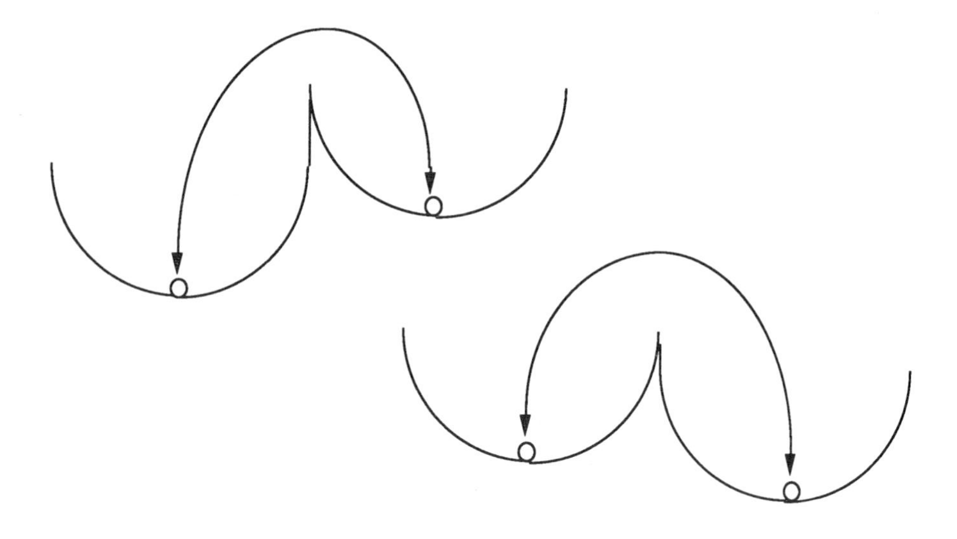

b. Second-order change (different level)

Figure 4.1 First-Order (a) and Second-Order changes (b)

1 See also Edward Yourdon, *Decline and Fall of the American Programmer* (Englewood Cliffs, Yourdon Press, 1992), Chapter 3 Peopleware; Frederick Herzberg, "One More Time: How Do You Motivate Employees?" *Harvard Business Review,* September-October 1987.

There are of course a multitude of reasons why change (of the second-order kind), cannot sometimes be introduced or at least introduced rapidly enough. I once consulted with a company who wanted to achieve ISO 9001 registration within six months; a very aggressive schedule. The company had a fairly good quality control and quality assurance program. As is often the case, the drive to achieve registration came from the V.P. of quality who had the support of other V.P.s but NOT all V.P.s. In this particular situation, although the V.P. of engineering was lukewarm to the idea of ISO registration, others within the department thought that ISO "would add another layer of bureaucracy," which would impede product development. To understand some of the hurdles the company was facing, one must look at Figure 4.2. The company in question was engineering driven; that is, the president, himself an engineer, invariably saw the engineering—and only the engineering—point of view. Although communication did exist between departments it was essentially a one-way communication; from engineering to the others. To complicate matters further, the V.P. of quality, a mildly authoritarian man in his own right, wanted to have as much control as possible of the implementation process. He had very specific ideas as to what needed to be done and in fact had already included all of his "opinions" in a very well-written and organized unofficial (ISO) quality manual. Needless to say, a certain amount of "latent" tension was slowly building up waiting to be released at some later date. Perhaps aware of the potential difficulty, the company chose to hire the services of a consulting firm to help guide them through the documentation process.

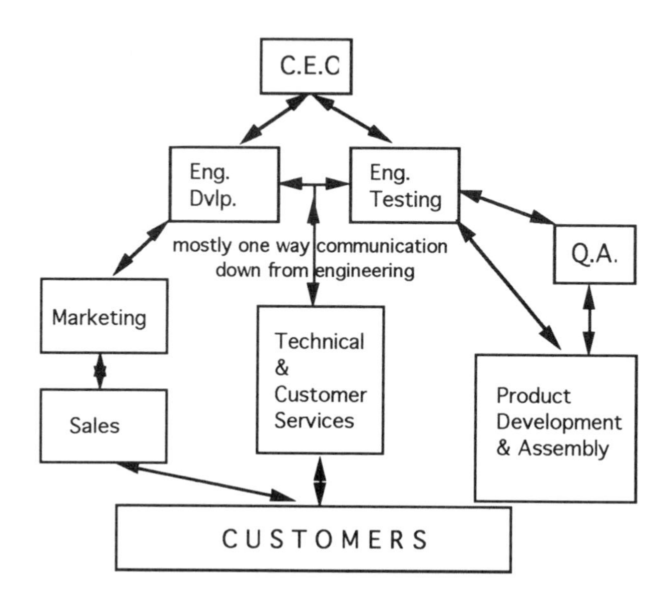

Figure 4.2 Simplified Organizational Flow

To believe that registration can be achieved without investing some time and effort which may or may not temporarily disturb the system, is simply unrealistic. Companies which have successfully achieved registration have recognized that in order to bring about second-order change, employees had to be informed and trained about the ISO implementation process. In addition, an adequate amount of time had to be allocated to organize, document and audit the quality assurance system in order to continuously assess its effectiveness and suitability. As Burleigh B. Gardner observed nearly fifty years ago in an ageless passage, "[E]mployees should have a feeling that the company's goal is worth their effort; they should feel themselves part of the company and take pride in their contribution to its goal. This means that the company's objectives must be such as to inspire confidence in the intentions of management and belief that each will get rewards and satisfactions by working for these objectives."[1] Companies that have recognized the need to introduce the paradigm shift suggested by

[1] Burleigh B. Gardner, *Human Relations in Industry* (Chicago: Urwin, 1945), p. 283.

Gardner and many others since have, generally speaking, benefitted more fully from the ISO 9000 experience.[1]

How to Bring About Second-Order Change

The road to ISO 9000 registration can not only be tortuous but also replete with seemingly insurmountable hurdles. To alleviate these difficulties and facilitate the advent of second order change, one must: 1) inform everyone as to the company's objective to achieve ISO 9000 registration and 2) delegate as much as possible the various assignments.

Well-focused and brief (one- to two-hour) training sessions are very effective in informing people as to the company's overall objective vis-a-vis ISO registration. These sessions should provide for a brief overview of the *nature* of the activity, the *reason* for the activity and the *procedure*. In other words, one should explain:

- **What** is the ISO 9000 series (nature of the activity)
- **Why** the company needs to achieve registration (reason for the activity) and
- **How** the activities will be carried on (procedures)

Explaining *what* and *why* are relatively easy. Formulating *how* is much more difficult and may require external assistance. Indeed, "How do you intend to find the time to prepare for registration?" is the most often asked question, which all too often remains unanswered.

---

1 The term paradigm shift was first introduced by Thomas Kuhn in his *The Structure of Scientific Revolutions* (Chicago: University of Chicago Press, 1970, 2nd ed.). The need to adopt a new methodology to better explain/fit events has been adopted in recent years by many business consultants and writers alike, who refer to the phenomenon as corporate culture change. See for example Jagdish Parikh, *Managing Your Self* (Oxford: Basil Blackwell Ltd., 1991, Chapter 13: "Doing Business in the New Paradigm", and Paul Watzlawick's concept of reframing, in *Change:*, . . op. cit., Chapter 8: "The Gentle Art of Reframing."

To explain *how* an organization will achieve ISO readiness, one must first determine *what* system(s) are currently in place; hence, the need for a pre-assessment visit which usually lasts anywhere from two to five days. These pre-assessment visits are usually subcontracted to a third party to ensure an unbiased estimation. Once the assessment is completed, one must next outline how the tasks will be undertaken.

The most effective strategy to adopt is to delegate the various responsibilities to as many people as possible. Naturally, this will require careful monitoring and a precise definition of what needs to be done, by whom and by when. Assuming that you have spent some time explaining the "*what and why*" of ISO (prior to implementation), you should be able to limit the unavoidable resistance to change brought on be the fear of the unknown. If you then allow for a period of adjustment and if you persist in re-stating and re-emphasing your (ISO) objectives, you should successfully bring about the changes required by the various ISO standards within eight to fourteen months.[1]

Bringing About Change

Based on the many books available on the subject of change, one should not be surprised to learn that there are in fact numerous theories about change. Edgar H. Schein traces some of the earlier theories of change back to Kurt Lewin's work published in the late forties and early nineteen fifties. Based on the plethora of research conducted over the past fifty years, Schein proposes that the following assumptions underlie his theory of change:

> 1. Any change process involves not only learning something new, but *unlearning* something that is already present and possibly well integrated into personality and social relationship of the individual.

---

[1] For a practical guide on how to bring about change see *Bringing Out The Best In Your People*. Bureau of Business Practice (Waterford, CT: Prentice Hall, 1992).

> 2. No change will occur unless there is motivation to change, . . . the induction of motivation is often the most difficult part of the change process.
> 3. Organizational changes such as new structures, processes, reward systems, and so on occur only through individual changes in key members of the organization; hence organizational change is always mediated through individual changes.
> 4. Most adult change involves attitudes, values, and self-images, and the unlearning of present responses in these areas is initially *inherently* painful and threatening.
> 5. Change is a multistage cycle . . . and all stages must be negotiated somehow or other before a stable change can be said to have taken place.[1]

Recognizing that change is a multistage process, it has been suggested that change can effectively be brought about either via the intervention of a consultant who can help the system improve its inherent capacity to cope with change; or via appropriate leadership.

On Leadership

Not surprisingly, there are as many books on leadership as there are about how to bring about change. Anyone who has gone through the business section of most bookstores might have come across such books as *The Tao of Leadership*, or some similar title. I do not intend to review all of the theories relating to leadership; to do so would require several chapters. I will limit myself to a brief overview of two models which are useful within the context of ISO 9000 implementation.

Hersey and Blanchard, for example, recognize four styles of leadership which they label *Delegating, Participating, Selling and Telling.* Each

---

[1] Edgar H. Schein, *Organizational Psychology* (Prentice-Hall, Inc., Englewood Cliffs, NJ, 1980), pp. 243-244.

style in turn depends on 1) the *maturity* of the subordinates, defined as their readiness to undertake a task, 2) the nature of the task (high or low) and, 3) the amount of social relationship required to achieve said task.[1] Thus, as the subordinates become more ready to accept the task (e.g., ISO 9000 implementation), the style of leadership should evolve from *Telling* (high-task and low-relationship behavior) to *Selling* (high-task and high-relationship behavior) to *Participating* (low-task and high-relationship behavior) and *Delegating* (low-task and low relationship behavior).

Other, such as Argyris and Schon do not believe that one can convince managers to become more participative by simply *showing* them how to become more participative.[2] Argyris believes that the basic cultural values (of a corporation) must first be changed before changes can be seen in individuals. Therefore, although there is not one correct model to apply, each has its place and importance *depending* on the nature of the task at end. There are, however, some critical functions that must be provided by all leaders irrespective of their style:

- They must be able to determine, articulate and transmit the basic goals or tasks to be accomplished.
- They must monitor progress towards set goals.
- *They must supply whatever is needed or missing for the successful completion of said tasks or goals.*[3]

This last requirement is often missing during the implementation phase of many so-called "ISO 9000 programs." When it comes to implementing ISO 9000, a blend of several leadership models may well be the best strategy to apply. In the early stages, the leadership might well have to be *autocratic* and *tell* subordinates that the

---

[1] P. Hersey and H.H. Blanchard. *Management of Organizational Behavior* (Englewood Cliffs: Prentice-Hall, Inc., 1977). pp. 160-165.
[2] See C. Argyris and D.A. Schon. *Organizational Learning: A Theory of Action Perspective* (Reading, MA: Addison-Wesley, 1978).
[3] Modified from Schein, Op. Cit., pp. 134-135.

organization *will* achieve ISO 9000 registration. One must, however, distinguish between an absolute autocrat and an enlightened autocrat. An absolute autocrat will irresponsibly demand that his/her company achieve registration within four months. An enlightened autocrat will not only recognize that the request to achieve registration within such a short period of time (four months), is not only very difficult if not impossible to achieve for most firms, but will also lead to some severe stress within most departments.[1] He or she will also recognize that (s)he must *sell* the idea to his subordinates by convincing them of the marketing/economic value or simply, the "total quality cum competitiveness argument." Having done so, our leader will next have to move on to the next stage and *consult* with his subordinates to determine *how* the task is to be completed and by when. Finally, after appropriate training and education, he can *delegate* who will do what.

Conclusions—Putting All Together OR "Where Do We Start?"

I recently was asked an interesting question: "Which paragraphs should we first address as we go through the implementation process?" I found the question interesting because I had never really thought of sequencing the implementation task on a per-paragraph basis. However when I starting to think about my answer, it occurred to me that I was in fact following a "logical" sequence which sort of marched through the standards paragraph by paragraph. There is in fact a *preferred* sequence—I hesitate to call it a logical sequence, for I am not sure it is necessarily so. Looking at the ISO 9001 standard, I would propose the following sequence which could conceptually be broken down into three major phases:

> **Phase I: 4.18 (Training)** It is amazing how many companies begin implementing one of the ISO 9000 standards *without* informing its employees as to what the series is all about! See below for further comments.

[1] The four- to five-month time span is not unheard of. A client of mine once told me that the electronic division of a Japanese conglomerate (electronic equipment, tapes, etc.) achieved ISO registration within five months!

Phase II: 4.1 (Management Responsibilities) This is perhaps self-evident but needs to be stated.

4.2 (The Quality System) Assess your current quality assurance system (if you have one in place). See below for further comments.

4.5 (Document Control) You should, very early in your program, decide what template or format to use for your document control process.

4.16 (Quality Records) Identify, per department, what is your quality record base.

Phase III: 4.17 (Internal Audits) Don't wait too long to formulate your internal audit process.

All other paragraphs, basically following the sequence: 4.3, 4.4, 4.6 etc. However, it is unlikely that you will implement phase III sequentially. Some of the clauses will have to be implemented simultaneously.

The reader should be warned that no matter what sequence, planning or logic is appied to the process, certain tasks must first be performed to facilitate success.

The implementation of any ISO 9000 quality assurance system consists of several tasks which can be broken down as follows:

THE TASKS

1. To compare your quality assurance system with the appropriate ISO standard
2. To close the gap(s) between the two systems
3. To formulate a methodology on how to achieve 2
4. To conduct internal audits to verify the effectiveness of the quality assurance system

HOW TO ACHIEVE ABOVE TASKS?

1. Need to communicate/inform/train people as to:

   a) *What* is the ISO 9000 series
   b) *Why* it is important to achieve registration
   c) State *objectives* and formulate *how* the company will achieve stated objectives (i.e. formulate commitment)
   d) *Delegate* tasks, which can only be achieved after a-c is first completed

HOWEVER:

• Training cannot be effective unless it is framed within the company's stated objectives,

AND

• One cannot delegate tasks unless:
   a) subordinates understand *why* it (i.e., task) must be done. This should not be the responsibility of the consultant, *management* should have a clear idea as to why ISO 9000 registration is required.
   b) the person delegating also understands *what* is involved and required.

THUS: To improve chances of success, upper management should first:

• Receive some training (one day), consisting of an overview of the ISO 9000 series, what is involved and which strategy to adopt

• Allocate some in-house training consisting of a two day seminar-workshop to a select team of "implementers"

- The team can then proceed to the TASKS

Having received proper training, the next step is to decide how to organize the quality assurance system, which is the subject of our next chapter.

5 The Quality Assurance System—Which Structure to Adopt?

The implementation of a quality assurance system designed to conform to either ISO 9001, 9002 or 9003 (ANSI/ASQC Q9001, Q9002 or Q9003) should be simple and straight forward. It essentially consists of one major objective:

- *Ensure that all documented requirements are practiced.*

Achieving that objective may, however, require anywhere from six to fourteen months. There are of course many strategies or methodologies available. Some companies delay their decision to "go for ISO" for several months and then decide to rush the implementation process. Others seem to prefer placing the cart before the horse by spending an inordinate amount of time deciding which registrar they should contact or hire *before* addressing implementation issues.

Such ineffective strategies usually extend the registration process and cause undue frustration. One of the reasons some companies delay the ISO registration process is because upper management has, perhaps unwisely, decided to first focus on the Malcolm Baldrige Award *prior* to achieving ISO registration. The rationale in such cases is "let us devote all of our energy towards the Baldrige and we shall then look into ISO." Such a strategy often leads to delays, frustrations and overall disappointment. Indeed, as is well known by many Baldrige applicants, spending thousands of man-hours preparing for the Baldrige does not guarantee receiving the award. One petrochemical firm which opted, in late 1989, for the "Baldrige now and ISO later" strategy still had neither as of late 1992!

Spending too much time focusing on "getting the right" registrar is also ineffective. It is certainly true that *some* registrars are booked for several months and therefore one should anticipate scheduling problems. Nonetheless, the scheduling of a registrar is unlikely to be the *first* concern for most companies. The first question to ask should be whether or not the company is ready to call in a registrar for a third-party audit. Most companies will discover that their quality assurance system will require, at a minimum, some reorganization. Others may well struggle with the all important task of deciding *how to organize* their quality assurance system.

Which Structure to Adopt?

The myriad of organizational permutations available to ISO 9000 users can be grouped into three categories schematized in Figures 5.0 a, b and c. For major corporations having either several business units or several sites, each consisting of major business units, the decision as to which organizational structure to adopt can be a most difficult one. Figure 5.0 a. illustrates one format whereby the corporation decides to produce a high level one (i.e., tier one) corporate quality manual which refers to lower tiered business units. Each business unit can in turn develop its own quality manual. As subsets of the corporate manual, these business unit quality manuals should be conceived as level 1.1 documents, that is, quality manuals within a quality manual. These manuals will in turn refer to subsets of documents known as tier two and tier three (see below for discussion on the Pyramid of Quality). There are advantages and disadvantages to using the organizational structure outlined in Figure 5.0 a. One of the advantages is that all documents can eventually refer to the corporate manual. The format should of course be standardized and that leads to one of the major difficulties inherent in that particular format. The amount of effort and coordination required to bring about a standardized and properly cross-referenced corporate quality assurance system across several divisions and business units is, at a minimum, enormous. Naturally, one has control over the amount of standardization and cross-reference required but nevertheless, one can expect spending countless hours trying to organize such a system.

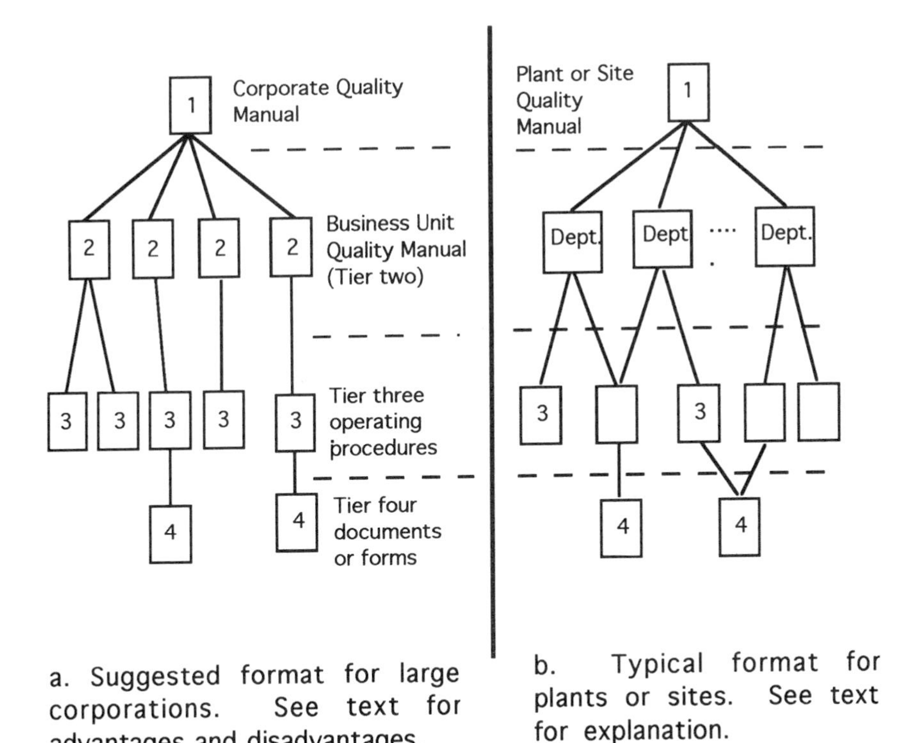

a. Suggested format for large corporations. See text for advantages and disadvantages.

b. Typical format for plants or sites. See text for explanation.

c. Possible format for small (< 50 employees) companies

Figure 5.0 a, b, c Suggested Organizational Structures

There is at least one more reason to shy away from the approach suggested in Figure 5.0 a., and that has to do with the registration itself. If a corporation seeks *one* corporate registration across several business units and geographical sites, the whole system is now interrelated. Thus, if *any of the units should fail the third party audit or any subsequent periodic audits, the whole corporation runs the risk of losing the certification.* Although the probability of losing a certification is small, the cost is indeed enormous if it should happen. Hence, the preferred format, which is suggested in Figure 5.0 b., where each plant, site or business unit organizes its own quality assurance system and applies for its own registration. In essence, Figure 5.0 b. is a subset of Figure 5.0 a.

Small companies may discover that they may not need an elaborate system. In some cases, distributors for example, all that might be needed is a quality manual and a combined tier two/three procedures. Irrespective of which organizational structure you may adopt, you will eventually have to address the very topic of implementation.

Which Strategy?

When first faced with the task of "implementing ISO", one of the most common "methodologies" (yet, not necessarily the most effective, see below), is to simply rephrase each ISO paragraph and follow the flow chart diagrammed in Figure 5.1. The reader could test his/her company readiness by selecting a clause with which (s)he is familiar and follow the flow chart. This strategy is not necessarily the best approach because it tends to ignore or bypass the all important question of how the quality assurance *system* will be structured to best fit the company's needs. An example will best illustrate what I mean.

The organizational structure depicted in Figure 5.2 is a schematic bubble diagram of a high tech company. Prior to focusing on what needed to be documented to fit the ISO model, it was first necessary to understand how the company was organized, how it functioned (i.e., product design and development, marketing, manufacturing or

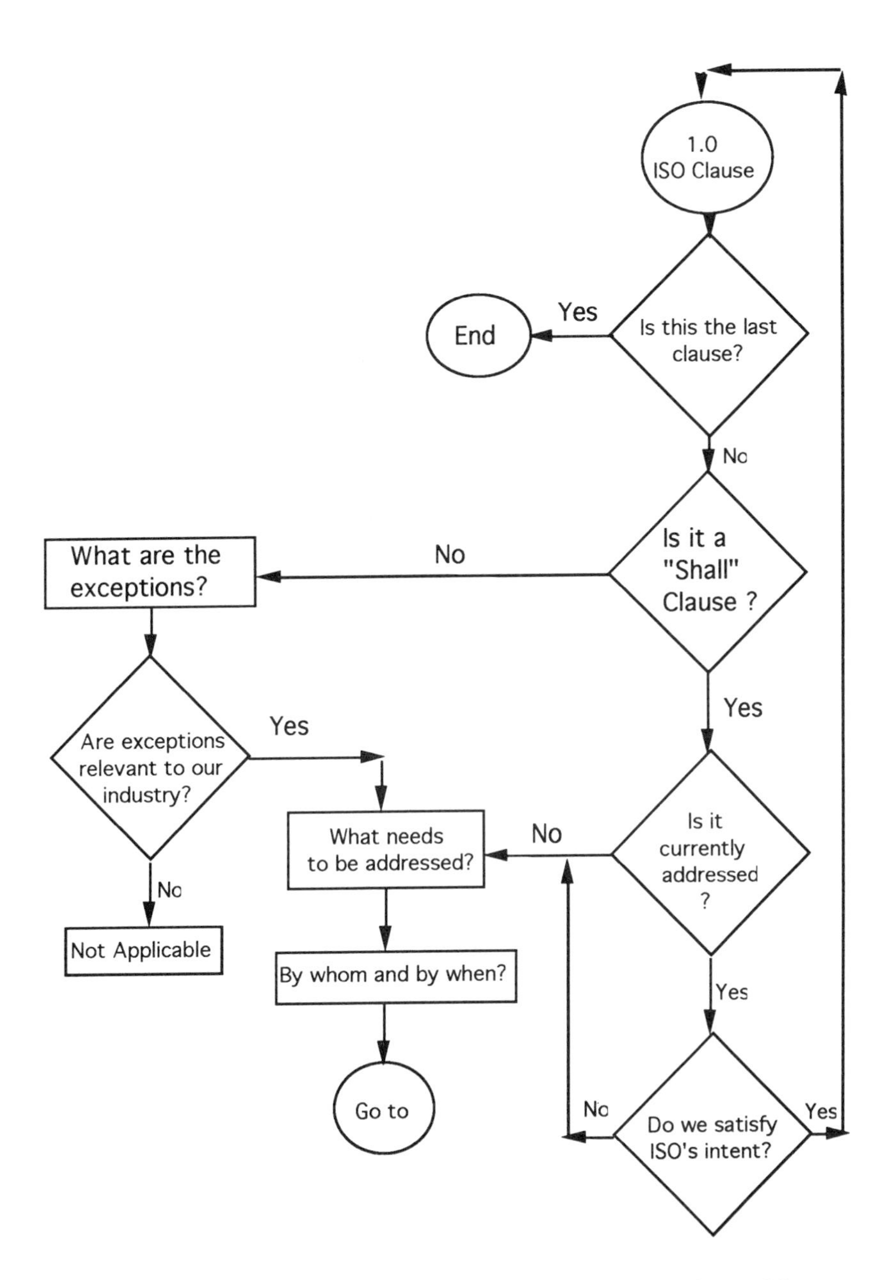

Figure 5.1 Implementation Flow Chart

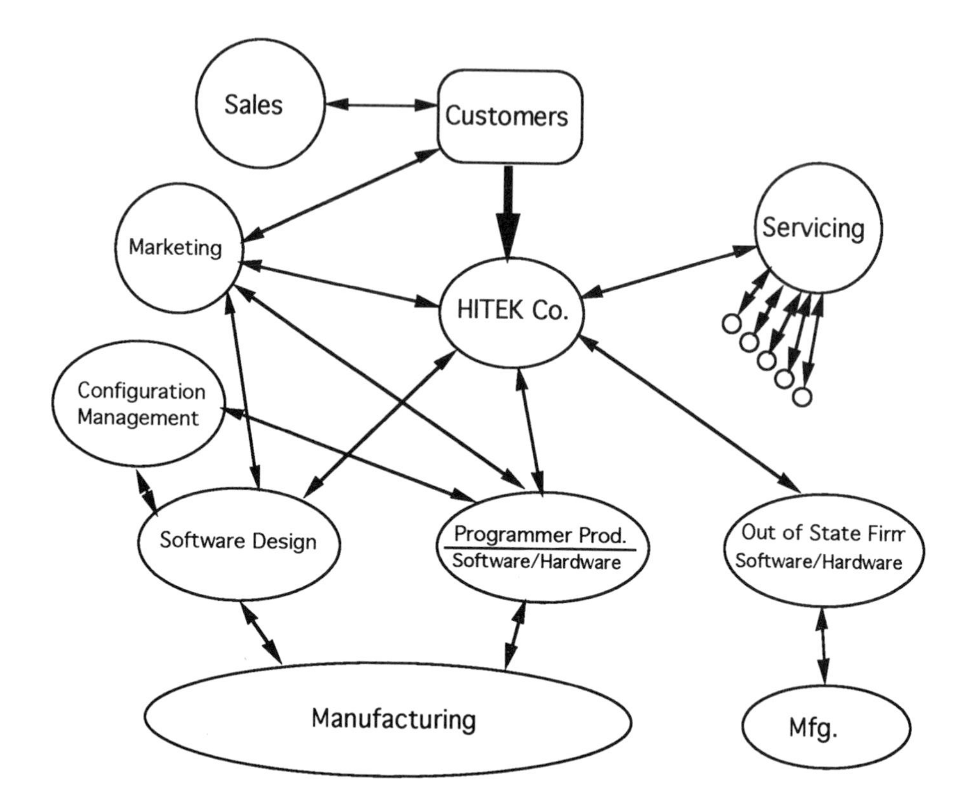

Figure 5.2 HITEK Organizational Structure

assembling, servicing, etc.), and how the various departments communicated with each other and with the outside world. As can be seen from Figure 5.2, the company in question consists of:

- Headquarters
- A marketing unit which develops a Marketing Research Plan (M.R.P)
- The MRP in turns "drives" a Product Development Document (P.D.D)

- The PDD is referred to by two independent divisions involved with design: software products and programer products
- Underneath the design divisions is a manufacturing/assembly unit which serves both divisions.

In addition, the following units can be found: 1) an independent site (with its own design department), recently purchased and located in another state, 2) a servicing center located in another state and "controlling" five other sites throughout the U.S., 3) overseas servicing sites and 4) other units such as sales, configuration management, technical function, maintenance, etc.

Having identified the *system* and all of its *interacting parts*, the next task was to 1) assess what documents were currently available, 2) determine if procedures were *still followed* or in need of rewrite and 3) decide which *format* shall be adopted. The standard format, favored by most, is the so-called *pyramid of quality* (Figure 5.3). The pyramid of quality consists of three to four tiers depending on how many details you wish to insert. Tier one (**required by registrars**) consists of the *quality manual*, a twenty five to thirty page controlled document (see ISO 9001 clause 4.5) structured much as the ISO standard itself (see Appendix D for an example).

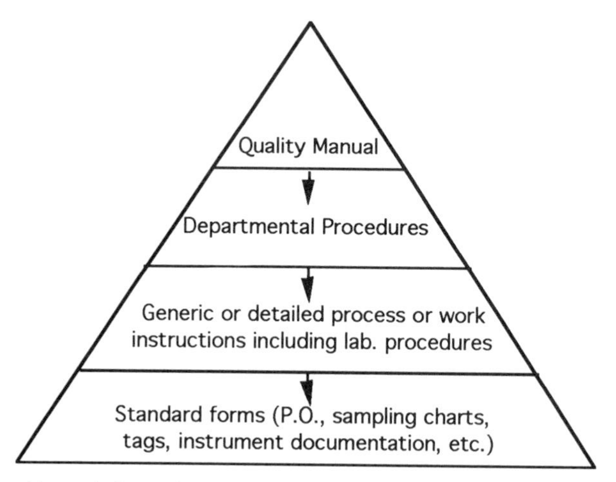

Figure 5.3 The Pyramid of Quality

Tier two (**optional**) expands on the contents of the quality manual. For example, whereas the quality manual would assure readers that nonconforming materials are reviewed for proper disposition (see nonconforming clause for exact wording), the tier two document would elaborate further on *how* this is achieved. Tier two documents can be structured in one of two ways: they can either be a brief document (one to four pages) describing the *department's functions* and *responsibilities* particularly as it relates to one *or more* ISO paragraphs, or they can simply be a procedure describing how a particular ISO clause is addressed within the company. Deciding whether or not tier two documents will be needed, how they will be organized and which departments are affected, can be most challenging.

Tier three documents explain how a process/procedure is followed. Examples of tier three documents would include: laboratory procedures, metrology procedures, operating instructions, generic design review procedures, generic or product specific process flow diagrams, testing procedures, sampling procedures, inspection procedures, internal audits checklist or related documents, packaging procedures, servicing manuals, etc.

With regard to the above mentioned high-tech company, a rather extensive document (the PDD), which included much valuable information directly relating to ISO requirements, had already been developed. Rather than re-invent the wheel, it was decided to refer to the PDD as much as possible (see Figure 5.4).

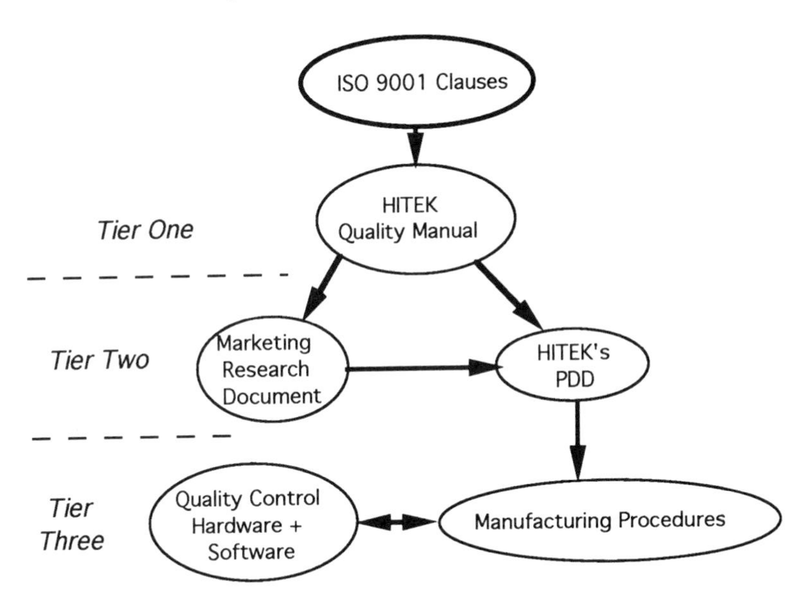

Figure 5.4 HITEK Document Structure

How to Organize Tier Two Documentation

Tier two documentation often is the most difficult tier to document because one does not really know what to include in it. Basically, the nature of the difference between tier one (quality manual) and tier two or three, is that of scope. The quality manual is an overall company-wide policy document which addresses *all* of the relevant ISO paragraphs. The scope of the tier two documents is more focused since it is designed to explain how a department or departments address a particular ISO clause or clauses. It is more general in scope from the tier three document which details how a process, described in tier two, is actually *performed.*

Before offering suggestions, it is important to recall that tier two documentation is not a requirement but rather a recommendation on how to organize a quality system. Tier two documentation is where you explain how the various processes relating to any of the ISO 9001, 9002 or 9003 paragraphs operate. The difficulties relate to the following questions: How much information/detail to include? What processes are we referring to? Who should write it?

The answer to these questions is not always straightforward. Let us look at the apparently simple process of contract review (4.3 in ISO 9001). It would seem easy enough to write a brief description of how the "contract review process" unfolds. However when the process is (verbally) described, one quickly discovers that it can be an involved process involving several parties. Figure 5.5 illustrates the point. In this particular situation, several departments participate in the contract review process. Who should then write the procedure? The department that coordinates all of the activities; in this case Marketing/Sales with the assistance and cooperation of Technical Services, R&D, Quality, Process Engineering, and the Laboratory. The tier two process should reference all of these individuals/departments since they all participate in the "contract review process." The contract review tier two document could of course reference other tier two or tier three procedures.

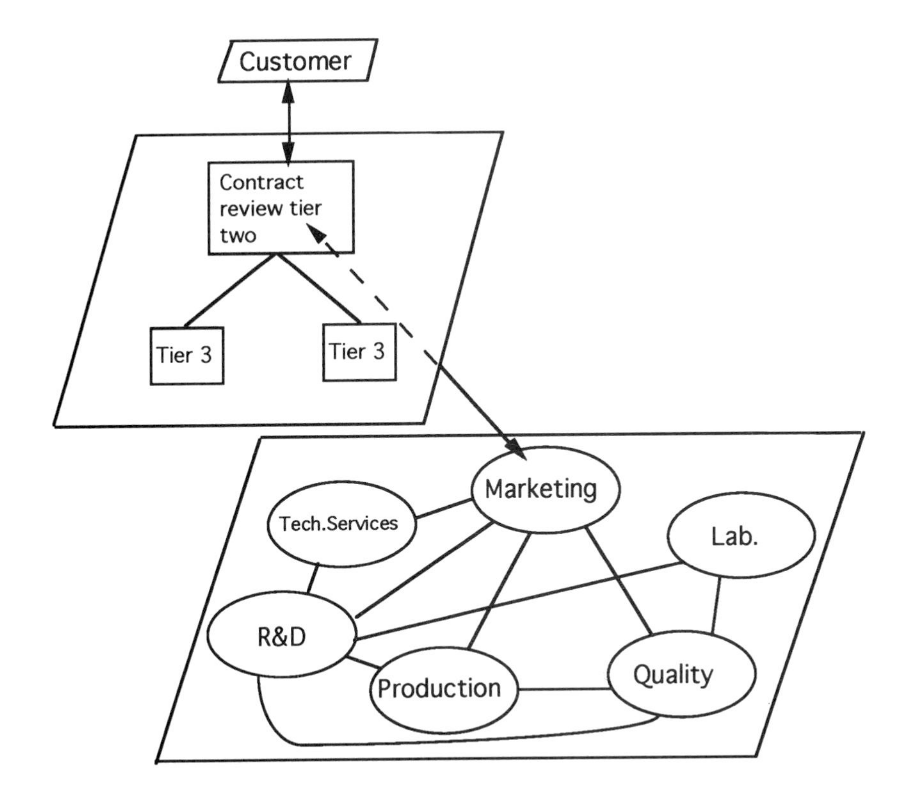

Figure 5.5 Tier Two Sample for Contract Review

To the question "How much detail should I include?", I do not have a specific answer except to suggest "as much as is required!" Tier two should not require more than two to four pages, particularly if you use flowcharts. Naturally, everyone involved in the process should have an opportunity to review the procedure. As for format, follow the same format that you use for the quality manual (see Appendix D for an example).

If customers order from a catalogue, there might not even be a need for a contract review tier two procedures. The procedure could be briefly stated in the tier one (quality manual). One might directly skip to a tier three to explain how an order is processed and verified. In fact, many companies inadvertently incorporate some or all of their tier two procedures in the quality manual. In such cases of course, the quality manual tends to be longer than the usual twenty five to thirty pages.

Robert Pannela's description of a calibration system provides a good model on how one could structure a tier two and tier three (see below) metrology document. Pannella suggests that the basic element of a calibration system should include:

- Organizational structure (Tier two)
- Responsibilities (Tier two)
- Planning process (Tier two)
- Environmental controls (Tier two or three)
- Maintenance of policies and procedures (Tier two)
- Interval of calibration (Tier two or tier three)
- Calibration procedures (Tier three)
- Adequacy of measurement standards (Tier three)
- Out-of-tolerance conditions (Tier three)
- Calibration sources (Tier three)
- Application of records (Tier four)
- Calibration status (Tier four, i.e., tags)
- Control of subcontractor calibration (Tier two)
- Storage and handling (Tier two?)[1]

When looking into tier two documentation, remember that some clauses are more likely to apply across the company, whereas other clauses are more department specific. It is difficult to say which clause fits where within a company simply because everyone does things a little differently. The *example* provided in Figure 5.6 illustrates how the concept of internal customer(s)/supplier(s) could be included within the tier two documentation. Flowcharts and the

---

[1] For further details, see C. Robert Pannella, *Managing the Metrology System*. (Milwaukee: ASQC Press, 1991), pp. 15-16.

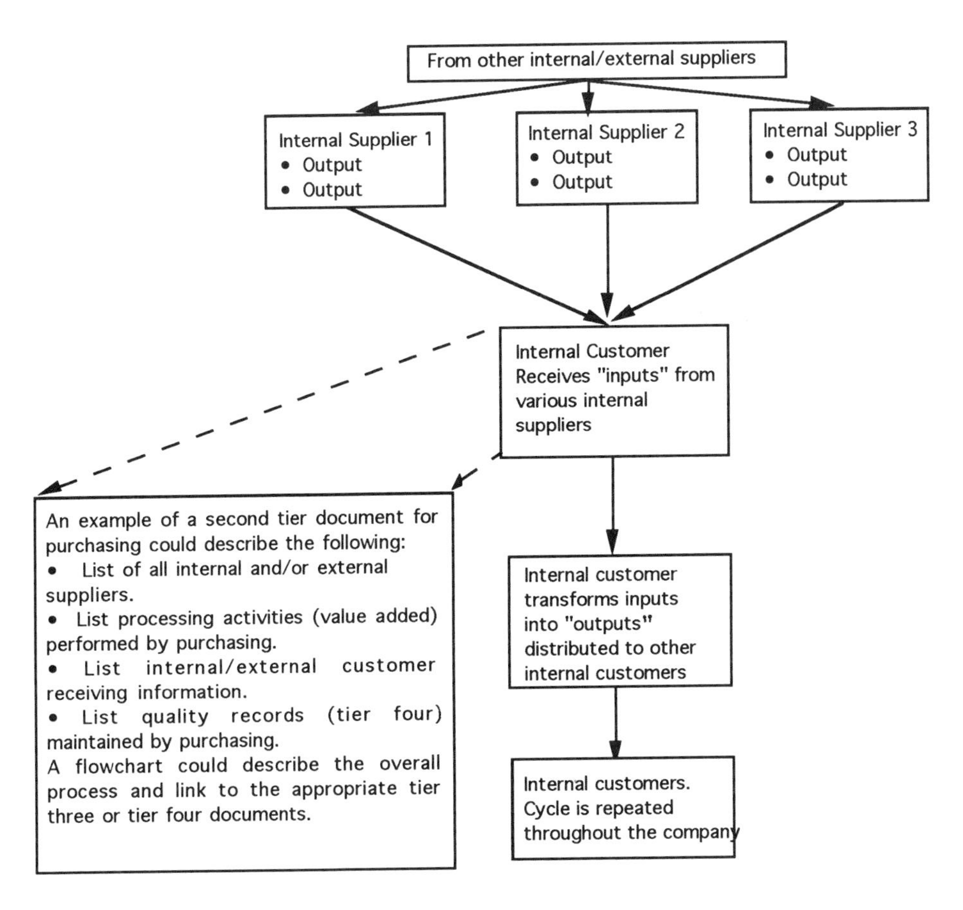

Figure 5.6 Suggestions on How to Structure Tier Two

hierarchical structure presented in Figure 5.5 could also be included in tier two documentation.[1]

Once every department has produced its tier two document, a master list could be prepared and stored either in the company's library, the quality manager/director office or some other convenient location. The location and control of tier two documents is up to the user.

---

[1] Examples of tier three documents are provided in my *ISO 9000: Preparing for Registration*. You should write third tier procedure to suit your needs, avoid too much detail (unless required).

Having established the overall structure of the quality assurance system, the next task is to organize the implementation effort.

Suggestions on How to Bring About Implementation

Prior to investing a significant amount of resources into the implementation effort, it is imperative that key decision makers be informed as to what will be required to successfully achieve registration within the expected timeframe. The information should include:

- √ A description of what needs to be done in terms of documentation.
- √ Estimated man-hours by departments needed to achieve above task. For firms up to 1,200 employees with an outdated quality assurance system, allow for approximately 2,200 man-hours including 30-50 hours of internal audits.
- √ Estimated completion dates for various phases such as: first draft review, first internal audit, etc.
- √ Emphasize that the system must be in place for approximately four to six months (some registrars are not so demanding).
- √ Estimated date for the official third-party audit. It is sometimes helpful to set the audit date and work backwards.

Once the above checklist has been presented to upper management it is imperative to obtain a commitment specifying the type of support management is willing to approve to ensure success within said time frame.[1]

---

[1] See also James Lamprecht, *ISO 9000: Preparing for Registration*. (New York: ASQC Quality Press and Marcel Dekker: 1992), Chapter 9.

It is usually more efficient and expedient to assign one person in charge of monitoring the various tasks. This person should in turn *delegate* responsibilities to key "implementers." Since "[D]elegation is nothing more than accomplishing results through the efforts of others. It is the manager's most basic and important tool." James Jenks and John Kelly, from whom the above quotation is taken, explain that "[P]oor information flow often characterizes the organization with delegation problems. Secrecy always inhibits effective delegation . . . Nobody wants to act without something in writing from the boss. A memo blizzard develops. Orders are passed down, but they aren't timely. More problems develop. The process starts over again. Poor delegation has been institutionalized."[1]

The delegation of tasks is certainly necessary to ensure success; however, it is not sufficient. The author has worked with companies where well-defined tasks were clearly delegated to so-called "implementers" who had committed to the tasks; and yet, results were not forthcoming. Internal leadership appeared to have been the missing ingredient.

The delegation of action items can be formatted as shown in Table 5.1. Once completed the list should be distributed to all concerned, *including upper management*, to monitor progress, or lack of. Table 5.1 is but a sample and includes only four ISO clauses. Naturally, all clauses would have to be addressed. Some tables include an additional column identifying *who* is/are responsible for the clause and *when* the task would be completed.

For small companies (i.e., less than 50 employees), a simpler strategy could be adopted. The following example illustrates how a fastener distributor outlined the various tasks.

---

[1] James M. Jenks and John N. Kelly, *Don't Do, Delegate!* (New York: Ballantine Books, 1985), pp. 6, 21.

• Each department was identified and the associated ***ISO 9002*** paragraph was written next to it.

DELIVERY (4.14.5, 4.7 ISO 9002)
SALES (4.3, 4.12, 4.13)
PURCHASING (4.3, 4.5, 4.7)
RECEIVING (4.7, 4.9, 4.12, 4.13)
INVENTORY (4.7, 4.9.3)
SHIPPING (4.7, 4.14.2, 4.12.3)
QUALITY CONTROL (4.3, 4.9.3, 4.11, 4.12)
MACHINE SHOP (4.8?)

Table 5.1 Sample ISO 9001 Action Item List

ISO Clause	*What* Needs to Be Done	*How* Is It Addressed & *Who* will Address It and by *When*?
4.1 Mng. Respons-ibility	• Shall define and *document* quality policy and *ensure* that policy is understood and implemented • Shall identify in-house verification requirements and provide trained personnel • Shall appoint management representative • Shall review quality system	• Mission statement • Internal audits procedures (trained auditors), quality council, staff meetings • Who? • Frequency and at what level? (Next staff meeting).
4.2 Quality System	• Shall have documented quality system in accordance with standard.	• Controlled via central computer • Implement via SOPs • Decentralized responsibility • Each employee is responsible
4.4 Doc. Control	• Shall maintain procedures to control all documents that relate to this standard • Shall review and approve documents by authorized person • Shall keep pertinent issues of documents available for effective function of quality system • Shall remove obsolete documents • Shall establish a listing • Shall re-issue after a practical number of revisions	• Each department will maintain procedures for document control • Department mgr. or designee will approve • A department designee will be responsible for controlling each document • Cast product manuals and chemical specifications are maintained at headquarters
4.8 Process Control	• Where applicable and if it directly affects quality . . . supplier shall ensure processes are carried out under controlled conditions • Documented work instructions. • Monitoring and control of process • Approval of process and equipment • Criteria for workmanship	• Process control have been established • Each department is responsible for the development and implementation of process control procedures

The following paragraphs/topics were then identified as having to be addressed:

- Quality Policy (4.1.1)
- Organizational chart (4.1.2)
- Verification (4.1.2.2)
- ISO representative (4.1.2.3)
- Mng. review (4.1.3)
- Document control (4.4)
- Subcontracted items (4.10)
- Training (4.17)
- Statistical techniques (4.18, Not yet applicable)

Since this particular supplier had to abide by the regulations set forth in Public Law 101-592 (Fastener Quality Act), the task was not too overwhelming. For example, requirements specified under paragraph 4.10 (ISO 9002) *Inspection, Measuring, and Test Equipment*, were subcontracted to an approved (A2LA) torque testing laboratory. An acceptance sampling plan (which needed some clarification) was used to submit a sample of approximately five to eight certified fasteners to the laboratory for testing. This procedure, which called for the testing of certified fasteners, was used as an added precaution. Records of all tests were fully documented.

Once all tasks have been defined, one still needs to develop a strategy on how changes will be introduced. For some companies, this may well be the most difficult task.

Conclusions

No amount of preparation and organization can help motivate uninformed employees. If tasks are simply delegated without explaining their purpose, it is very likely that procrastination will ensue. No one will have the time, or claim to have the time to write or modify a particular procedure. Unfortunately, delays can still occur even after in-house seminars are offered to a group of twenty to thirty implementers. Obviously, in such cases the need to change is either not accepted, not understood or, worse yet, not believed. If employees perceive the efforts required to achieve ISO registration as one more temporary program designed to either reinforce status quo habits or merely "window dress" inefficiencies, they will resist these attempts to introduce what has been described as first-order change. If genuine second-order change needs to be introduced, it will have to be introduced *at all levels* of the organization.

6 Sample Documentation

> "There is a wealth of evidence to indicate that concentration upon the review, analysis, and improvement of white collar procedures is a profitable undertaking in terms of money saved, of emproved employee morale, and of customer satisfaction." William A. Gill.[1]

It has been my experience that, for the majority of companies, the most difficult task to undertake is the writing of the tier two documents. Yet, even though tier two documentation is optional, it is an important set of documents which should be designed to capture the intricate interdepartmental flow of information and tasks within a company.

The following examples are presented to assist companies develop their documentation. I should emphasize that the examples are an amalgamation from several sources. The text and format is presented only as a model to assist others. As with any model, the suggestions presented here are meant to be guidelines and readers are encouraged to borrow, modify, adjust the model so as to develop their own set of documents. There are in fact two models: the first model relies on a generic quality manual, which usually reads like a policy manual (see Appendix D for some examples). The tier one manual may in turn refer to particular SOPs (standard or *suggested* operating procedures) of which there might be hundreds.[2] The second model, presented here, favors a different approach whereby the tier two (departmental) documents go beyond mere description of procedures.

The interrelationship between documents should follow the pattern illustrated in Figure 6.1. Ideally, each tier should refer to the tier

---

[1] William A. Gill, "Systems and Procedures," in Victor Lazzaro (editor), *Systems and Procedures* (Prentice-Hall, Inc., Englewood Cliffs, NJ, 1959), p. 4.

[2] The acronym SOP usually stands for standard operating procedure. However, I would like to propose the use of suggested operating procedure, simply because it implies that some procedures must sometimes rely on the intellingence and know-how (training) of its users.

below. Since tier one (quality manual) documentation has already been addressed in Appendix D, I will focus on the elusive tier two.

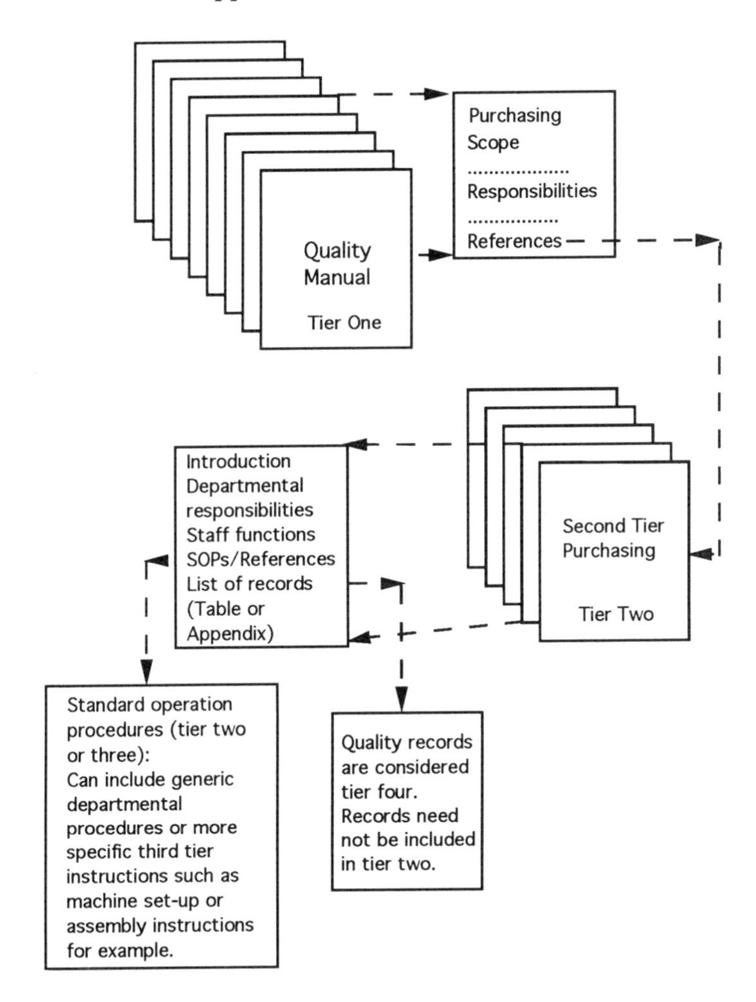

Figure 6.1 Structural Interrelationship Between Tiers (Focus on Purchasing)

Prior to developing a tier two document, each department manager or his/her delegates should ask the following questions:

- Who are my internal/external customers and suppliers?
- What value added process is added within the department (i.e., responsibilities) AND *how* (i.e., guidelines/SOPs/flowcharts) are these value added functions performed?
- What are the department's "deliverables" (also referred to as outputs?
- What departmental records are generated (see Figure 6.2)?

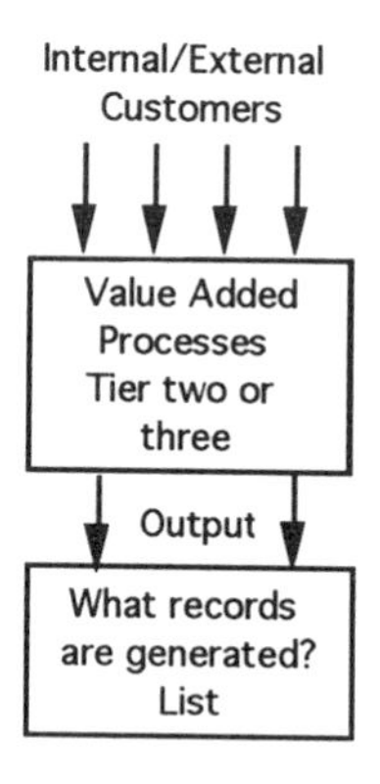

Figure 6.2 Value Added Process

If each department/function were to answer the above questions, most of tier two would be generated. Having answered all of the above questions the department's tier two document can then be written and structured to include the following headings:

- Introduction
- Organizational chart
- Staff responsibilities
- Operational guidelines (flowchart) or (SOPs)
- References (i.e., tier four quality records)

Let us review some examples. We start with a relatively easy clause, Purchasing. The following pages illustrate: 1) what to include within the quality manual page(s) and, 2) follows with a few pages from the tier two purchasing department.

5.0 Purchasing

5.1 General

The purchasing department ensures that all purchased materials and subcontracted services conform to specify requirements.

5.2 Assessment of Sub-Contractors

A list of approved suppliers is maintained by the purchasing department. These suppliers are selected on the basis of their ability to meet specific requirements including commercial, quality and reliability criteria. Selection criteria depends on the type of product being purchased and include historical performance as well as periodic appraisal conducted by the quality assurance department. Records of such appraisals are maintained by the director of quality assurance. The effectiveness of the appraisal system is reviewed annually.

5.3 Purchasing Data

Purchasing documents include a clear definition of the products and/or services required. Where applicable, purchase orders may include one or more of the following:

- A description of materials, drawings, specifications, test requirements or other technical data is required.

J.L.L. Inc.	**Quality Manual**	Page: 19 of 28 Issued: 9/12/92 Previous release: 8/17/92

- Identification and marking specifications as required by various safety certification agency.

- Requirements for shipping and packaging.

All orders are reviewed and approved prior to final release.

5.4 Verification of Purchased Product

This clause is not applicable since J.L.L Inc. does not purchase product from its customer for incorporation into its final products.

5.5 Responsibilities

See Purchasing Operating Document.

References

1) Purchasing Department Operating Document
2) Sub-Contractor Evaluation Survey

Written by: Purchasing Mngr. Approved by: Purchasing Mngr.

The next few pages are an example of how the Purchasing Operating Document (Tier Two) could be structured.

Purchasing Department

J.L.L Inc.	**Purchasing Department**	Page: 1 of --------------------------- Approved By: Purchasing Manager (signature)

0.0 Introduction

The following document defines the scope, responsibilities and standard operating procedures for the Purchasing department at J.L.L. Inc.

1.0 Organizational Chart

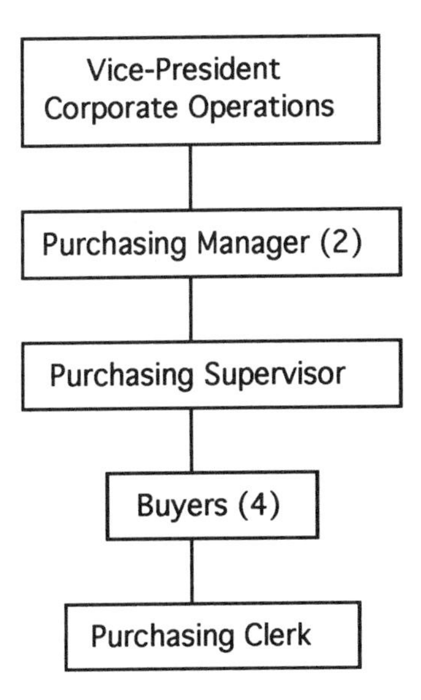

2.0 Departmental Responsibilities

Internal customers: Production, operation support, accounting, corporate research and development, customer service and technical services.

The purchasing department performs the following activities:

- Procurement of materials and services as demanded by the organization.
- Development of procurement strategies.

- Provide information on goods and services to the organization as a whole.
- Act as a liaison between the various departments and the external market place. Develop a long-term relationship between all parties.
- Solve problems with suppliers and internal customers to resolve discrepancies, poor performance or other issues.
- Evaluate suppliers (with the assistance of the Q.A. department), via audits, surveys and performance evaluation.
- Develop new alternatives or new buying sources.

3.0 Staff Responsibilities

3.1 Purchasing Manager

- Order placing to secure materials and services required by the organization.
- Negotiate or bid major requirements of supply in a cost effective manner. This include preparation and issuance of requests, analysis of offers, selection of suppliers and finalization of contracts.
- Development of procurement strategies including market analysis.
- Provide specific information as required.
- Work with suppliers/vendors to resolve discrepancies, solve problems or improve performance.
- Provide procurement services to support the needs of the production, maintenance and other service groups.
- Obtain prices and deliveries for goods and services through competitive bidding and/or negotiation when adequate specifications are available.

- Issue purchasing orders and ensure that goods and services are received on a timely basis and are of the expected/specified quality.
- Interview prospective new vendors.
- Supervise and lend assistance to subordinate.

3.2 Purchasing Supervisor

- Solicit requests for quotations and analyze data after receipt of quotes.
- Purchase goods and/or services required in a cost effective manner.
- Develop supplier partnership.
- Expedite orders.
- Conduct supplier audit and surveys.
- Negotiate with suppliers.

3.3 Buyers (4)

- Order placing to secure materials and/or services.
- Negotiate or bid materials and service.
- Analyze offers, select suppliers and finalize purchase or rental.
- Provide information to internal suppliers as required.
- Work with vendors to resolve invoice and receiver discrepancies.
- Maintain user manual on current software and work with MIS group.
- Work with internal customers to coordinate return of merchandise for replacement.

3.4 Purchasing Clerk

- Secretarial and general clerical duties for the department.
- Types correspondence, contracts, rental agreements, and quotations.
- Enter data in computer.

Purchasing Department Operating Procedures

The following procedures are meant to be general guidelines. When in doubt as to how to interpret these guidelines, please contact the purchasing manager. The following procedures are listed:

Alternate Source Approval
Contract Distribution
Corrective Action
Record and Filing Retention
Invoice Reconciliation
Material Specification Procedure
Supplier Evaluation and Selection
Request for Shipping

[*Note: A sample administrative procedure is included for reference. Figure 6.3 is a sample Process Flow Sheet that could be used for administrative flow charting (see Figure 6.4 for further details). Once the process is detailed, a flow chart could be produced by simply linking all of the symbols. The Flow Sheet is derived from forms used for time study analyses developed decades ago. Notice that the use of the decision symbol ◆ is not included here only because I did not want to include a process that would require branching. The decision symbol is nonetheless often required for a process flow diagram. The reader should work on his/her own third level documentation. The use of very detailed and/or very specific procedures is not recommended unless absolutely required (i.e., laboratory or perhaps assembly procedures).*

The more detailed a procedure the more likely an auditor is likely to find some deviations and hence nonconformances.[1] *Whenever possible, approach administrative procedures as guidelines. Rather than calling such procedures Standard Operating Procedures, one should perhaps call them Suggested Operating Procedures. It would be a good idea to have these SOPs under document control.*]

References

Reference any pertinent fourth tier records such as purchase order, corrective action forms, supplier surveys, purchase requisition, contract log, material specification form, etc.

[1] As Harvey J. Brightman explains, "Unless the use of standard operating procedures (SOPs) mandate innovative behavior as an organizational standard, innovation is unlikely to occur. [Moreover] [n]ovel problems require novel solutions. SOPs, while useful in dealing with well-structured problems, generally fail to solve ill-structured problems." From *Problem Solving: A Logical and Creative Approach* (Business Publishing Division. College of Business Administration. Georgia State University, Atlanta, Georgia, 1980), pp. 32, 34.

PROCESS ANALYSIS CHART

Procedure Name:______________________________ Date:_______

Department:_________________ Approved by:_________________

Step #	Description of each step	☐ Operation	○ Inspection	► Flow	▽ Storage	D Delay

Figure 6.3 Process Flow Worksheet

PROCESS ANALYSIS CHART

Procedure Name: Alternate Source Approval   Date: 9/10/92

Department: Purchasing   Approved by: Purchasing manager

Step #	Description of each step	□ Operation	○ Inspection	► Flow	▽ Storage	D Delay
1	Analysis of justification to approve alternate source (cost, quality, etc.)	√				
2	Submit technical data* to appropriate dept. for evaluation.	√		⇒		
3	Notify supplier of the evaluation status.	√		⇒		
4	Approved suppliers must submit: 1. Name of self 2. Site of production 3. Product data (specification plus Materials Safety Data Sheet)* 4. Sampling procedure* 5. Test methods* 6. SPC if available*	√	O			
5	*Insert other steps as required.*					
	*Note: The ⇒ indicates a paper flow. A * indicates that records are generated.*					

Figure 6.4 Purchasing Sample Flow Sheet

Multi-Departmental Procedures

The above example illustrating how the purchasing department could organize its tier two documentation is relatively easy because in this instance the contract review paragraph (4.6 for ISO 9001) is pretty well confined to the purchasing department. This is not the case for most other (ISO 9001 or 9002) paragraphs which invariably involve several departments working with each other in order to achieve several objectives. In such cases, the various processes have to be broken down into several sub-processes. A case in point is the design function which, for most companies, is a crossfunctional interdepartmental series of activities. In most cases the design function is managed by one or more engineering department(s) which in turn (supposedly) interact with other departments. Depending on the type of industry and product, these departments might include a combination of some or all of the following: marketing/sales, manufacturing, quality assurance/control, customer service, technical services, and perhaps one more department. It has been my experience that in the majority of cases, the interrelationship between the above mentioned departments tends to be less than cordial. Everyone appreciate the constraints of the other party, yet each is convinced that there is the only one correct way to proceed—their way. In such cases, the implementation of an ISO 9001 or 9002 quality assurance program can be helpful if, *and only if*, each department is willing to identify their internal customers/suppliers, determine their needs and expectations and resolve their differences. I do not have any words of wisdom on how to bring about harmony. I can however suggest that each department begins by preparing its own tier two document following the structure presented for purchasing. If each department were to describe how it operates and how it interfaces with other departments, one would at least have an operational model which could be improved as needed.

Conclusion

There are several options available when documenting a quality assurance system. My personal collection of over a dozen quality manuals reveals that manuals are as diversified in style and contents as the person(s) who wrote them. Consequently, a useful strategy is to identify the structure of your current system, and proceed to build upon *and improve* that structure. One thing you should never do is to start teams on documentation writing *without* first assessing what you currently have, how outdated it is, how much updating will have to be done and how much will need to be written.

How you structure *your* system is basically up to you. Do not try to blindly copy someone else's quality assurance system. Of course, you should always try to review how others have done it and even read books on the subject. However, remember that your operating constraints are not necessarily the same as someone else's. I have discovered for example that many U.S. firms simply borrow, and/or copy verbatim the ISO 9000 quality assurance system implemented by their U.K. subsidiary. The rationale behind such actions, I suppose, is that time and money will be saved. In some cases, that might well be true but it can also be a costly venture—particularly if the U.K. plant has implemented an elaborate ISO 9000 edifice. The model suggested in this chapter is intended to be a guideline—feel free to improve and adapt as you see fit.

7 Internal Quality Audits

Internal quality audits are yet another company-wide activity required by ISO 9001 and 9002. This is the only paragraph which allows for assessing the *effectiveness* and continued *feasibility* of the quality assurance system. Internal audits should be conducted as soon as the documented quality assurance system is in place. One need not wait until the system is fully documented to conduct the first audit. In fact the audit team can conduct a couple of audits as the system is being implemented. There are several advantages for doing so. Firstly, it allows both parties—auditors and auditees—to interpret, question evaluate and analyze first hand the everyday practical issues surrounding the standard. Secondly, it allows the audit team to monitor the implementation progress of each department. Valuable lessons can be learned on how a particular department addressed a set of requirements. Thirdly, the auditors get to practice their auditing skills.

It has been the author's experience that during the early stages of internal auditing, most audit teams—having intimate knowledge of the processes and the weaknesses of the "working system"—tend to focus on uncovering as many nonconformances as possible. In some cases, some of the nonconformances are tangentially related to ISO, in other cases they tend to be of the "nit-picking—I got you" nature. Internal audits which merely focus on the policing aspect of auditing quickly lose their credibility and thus effectiveness. Conducted properly, however, first-party (i.e., internal) audits provide an excellent opportunity to *verify* whether or not the documented system is indeed implemented at all levels.

Internal audits should be:

- *Procedural* (see sample).

- *Planned and documented* to determine the effectiveness of the quality system.

(Note: Whereas ISO 9001 states that the "supplier shall carry out a comprehensive system of planned and documented internal quality audits . . .", ISO 9002 merely states that the supplier shall "carry out internal quality audits . . .". Nonetheless, even though 9002 does not require the supplier to document his internal audits documentation it is recommended.)

• *Scheduled* according to the status and importance of the activity.

This simply means that you have the option to audit each department as frequently as you wish. For example, you may wish to audit all affected departments twice a year (at least) or you may choose to audit your laboratory every quarter and the purchasing department once a year, for example. Although the frequency of audits is left to the supplier's discretion, it is not reasonable to suggest (as I once read in a quality manual), that "internal audits will be conducted when deemed necessary." Most companies plan internal audits at least twice a year.

In addition, audit findings shall be:

• *Documented* and brought to the attention of the appropriate management which shall take "timely corrective action."

When addressing Internal Quality Audits, it is important to remember Shewhart's **P**lan, **D**o, **C**heck, **A**ct cycle. The internal audit team must not only Plan/schedule the audit, it must also conduct the audit (Do), verify that the documented system matches the implemented system (Check) and finally and, most importantly, management must *Act* upon all nonconformances in a timely fashion. An audit status report with many "no reply" comments can indicate that management is not taking the internal audit process very seriously (see **Closure**).

Size of the Internal Audit Team

The size of the audit team will naturally vary with the size of the organization. Some companies make the mistake of assigning the internal audit function to only one person. Auditing is a demanding, if not tedious, activity. It is therefore imperative to share the audit responsibilities with more than one person. Failure to do so will quickly lead to burn-out. Moreover, the internal audit process is much too important to trust to only one person.

Many companies train anywhere from two to twenty people to become internal ISO auditors. There is really no right number or scientific formulas that one can apply to determine the correct number of internal auditors except to say that it must be greater than one. Small companies (up to 100-250 employees) should have at least two individuals in charge of audits (one lead auditor and one auditor).

How to Conduct Internal Audits

Assuming that each member of the internal audit team is a willing participant, the success of your internal audits will generally depend on:

- The amount of *preparation* and organization of the audit team
- The degree of *familiarity* with the standard and the company's quality system
- The *human relation* skills of each member
- The type of *reporting*
- The success rate of your *closures*

Preparation

Preparing for an internal audit usually does not require as much time as preparing for a third-party (independent/external) audit. This is simply due to the fact that even though the auditors do not audit their own department, they nonetheless have (or should have) a more in-depth understanding of their company than any external third-party auditor. Nonetheless, much preparation will be required, particularly in the early stages. The following activities are recommended:

- Ensure that the people you will be auditing are well aware of the various ISO requirements. The author has participated in some internal audits where the auditee(s) had never seen the ISO standard. It is difficult to comprehend how an organization could attempt to implement an ISO quality assurance system—and conduct internal audits—without informing the affected departments as to what is required. One easy and *partly* effective way to do so is to, at a minimum, distribute copies of the standard. I have emphasized "partly" because the standard still needs to be interpreted and applied to the company's needs. See Sample 2 for suggestions on how to address the above issues.

- Define the *purpose* and *scope* of your audit (see Sample 3). The scope should define what will be audited and to what standard or paragraph within a standard. There is no need, for example, to cite (except perhaps as a verbal observation) OSHA discrepancies during an ISO audit and vice versa.

- Estimate how much resources (time and manpower) will be required.

- Schedule the time and day of the audit (insure that there are no scheduling conflicts with the auditee(s), see Samples 1 and 2).

- Decide on which audit trail method(s) the team will adopt. A few options are available. You could:

  (1) follow the product downstream (i.e., from purchasing/marketing to receiving inspection *down* to shipping)

(2) reverse the process and move upstream

(3) plan a particular audit trail which would combine upstream and downstream auditing

(4) assign (ISO) paragraphs to team members and periodically meet to consult and share notes

(5) develop your own style which might include a combination of all of the above options

- Develop a checklist which will help you determine what questions to ask (see Table 7.1). What documents will you want to see? [**Note and caveat**: This is particularly important during your first two or three audits. **Please remember that the checklist is NOT a substitute for the audit. Nor is it a simple rephrasing of the ISO standard.** Some auditors, even professional third-party auditors, seem to be totally enamored with their checklist, never deviating from its contents. The checklist is nothing more than an *aide-mémoire* (i.e., a reminder of what needs to be done). The checklist sample listed in Table 7.1 is derived from the U.S. Air Force environmental audit checklist found in the Appendix of Lawrence B. Cahill's *Environmental Audits*. The book contains some excellent information on training auditors and managing and conducting (environmental) audits. The recommendations and advice apply equally well to the ISO 9000 audit process.

Table 7.1 Sample Audit Checklist for Statistical Techniques

Statistical Techniques	Comments	Rating
1. Are statistical techniques used? 2. (Optional) Is the use driven by customer requirements or in-house interest? 3. What techniques are used? (*Note: Several options are available: SPC, DoE, Tabulation, Frequency, etc.*) If statistical techniques are used, check for proper use. For example, some sample questions for SPC would include: a. Could you show me samples of your charts? Examine charts, and verify that: 1. The proper charts are used. 2. Control limits are correctly calculated. 3. How often are charts updated? Sampling plan? 4. Check Cpk ratios or other capability measures. The above tasks can be arrived at by simply asking questions and letting the interviewee demonstrate his/her competence. For example: Why did you use a C chart? Could you show me and explain your capability studies? The same could be done for DoE studies, assuming of course the auditor has some knowledge of DoE.	Include here your observations. I have selected the statistical techniques paragraph because it is relatively easy to audit (assuming you have a good knowledge of statistics), and it is an optional paragraph. Indeed, remember that paragraph (4.20) does not require the "supplier" to use statistical techniques. If statistical techniques are used (often driven by customer requests), then all you can do is verify that the techniques are used correctly. If you should uncover misapplications of a technique (not too difficult to do), all you can do is note the misuse. For other clauses, many registrars simply rephrase the standard and append a "yes" ,"no", "NA" column. I do not much care for such simplistic checklists. They transform the ISO 9000 audit process into a regulatory exercise (see EPA audits for example). Nonetheless, such checklists may be valuable for training and can be used by inexperienced auditors.	Should be rather simple: "O.K., Needs improvement, or Not applicable."

Familiarity with the Standard

No amount of preparation will help you if you are not familiar with the content of the ISO standard. Unfortunately, in-depth knowledge of the standard can only be acquired through repeated readings and application of the standard. Consequently, do not be surprised or frustrated if, during your first couple of internal audits, you will feel somewhat uneasy, not sure of what to ask, or look for. This is a natural process. Hence the importance in the early stages of a well prepared *audit plan* and a *checklist of questions* (just in case you don't know what to ask for). During the first couple of audits, I would recommend that two or three internal auditors audit the same department. The advantage of doing so is that it allows each member of the audit team to think of the next question while a fellow team member is pursuing a line of questioning. Be careful not to "fire" too many questions simultaneously at the auditee. Take turns and be patient. Finally, do not feel compelled to search until you have found what you consider to be an appropriate quota of nonconformances. In the majority of cases you should not have to search too long before nonconformances are uncovered or volunteered by the auditee.

Internal audit teams conducting pre-assessment audits may be faced with yet another difficulty—the absence of a fully documented quality assurance system. Nonetheless, although it might be difficult to develop a relevant checklist without the help of a quality manual or other pertinent documents, such checklists can be written if need be. When auditing an undocumented system, auditors are faced with the additional challenge of literally discovering/understanding the system as they simultaneously audit it! When faced with such a situation I would recommend that the services of a consultant might be the most efficient way to assess "ISO readiness."

Human Relations

Any auditor must learn to develop some human relations skills. The purpose of the audit is, in part, to insure that the documented system matches the actual (working) system. However, this exercise does not imply that an auditor must find as many nonconformances as possible or that (s)he must audit until at least one nonconformance is found. One of the main functions of the auditor is to *assist* in the continued application/implementation *and* improvement of the quality assurance system. To achieve that goal, the auditor *must* learn to *listen* and get *input* from people on how the system can be improved. As Nancy W. Girvin explains, ". . . it is **people** we audit (not the requirements), **people** who get our report (not the findings), and **people** who make effective corrective action happen (not the audit)."[1]

Reporting

Once you have completed your audit, you should conduct an exit interview which is then followed by a formal audit report. During the exit interview you will need to present and, if need be, discuss preliminary findings. As an auditor, your objective is to ensure that your findings are as concrete as possible. You should be able to state:

- The nature of the nonconformance.
- Where the nonconformance occurred (department, assembly line, person interviewed, etc.)
- The nature of the nonconformance, that is, whether it is a written or verbal evidence.
- When the nonconformance occurred (date, some even include the time).
- Which ISO paragraph applies. This apparently trite observation is worth stating simply because some auditors tend to forget that they are auditing to an ISO standard.

---

[1] Nancy W. Girvin, *Writing Audit Findings: Be Reasonable!* ASQC Quality Congress Transactions-Nashville, 1992, p. 860.

- Which section of the company quality manual or other tiers applies.

During your preliminary presentation and final report, you should avoid including long lists of identical or nearly identical nonconformances. Rather, you should attempt to *cluster* similar deficiencies as elements of one global finding. There are a couple of advantages in doing so. First of all, the reporting of cluster findings demonstrate that the auditor has a good command of the auditing process and, more importantly, that the overall system has been evaluated. Secondly, cluster findings are more likely to be accepted simply because, unlike the old fashioned nitpicking deficiencies, they are perceived to be more reasonable. I should, however, warn that what is considered nitpicking in one industry (e.g., manufacturing of foam used for packaging) might be considered of critical importance in another (e.g., pharmaceutical or aerospace industries).

If everyone agrees with the nonconformances, the next step is to propose a schedule as to when and how the nonconformances will be resolved.

Closure

This follow-up activity is occasionally ignored, not by the audit team, but by the department responsible for addressing the corrective actions. Closure can not be achieved until corrective actions are implemented (see Sample 5).

Samples

The examples listed below consist of sample letters detailing how to:

- *Schedule audits (Sample 1)*
- *Inform auditee of intended audit (Sample 2)*
- *Audit report format (Sample 3)*

- *Audit evaluation questionnaire (Sample 4)*
- *Audit finding response form (Sample 5)*
- *Audit Status Report (Sample 6)*

These samples should provide the reader with enough guidelines on how to address the ISO internal quality audit clause.

Sample 1

It is always a good idea to offer advance warning as to when each department will be audited. The following pages help illustrate how to schedule audits.

Distribution:

Name	Department
Steve Mickey	Product Eng.
Robert Hall	Metrology
Allan Moor	Purchasing
Joyce Hornbeck	Quality Assurance
Bruce Middleton	Training
Jack Finn	Manufacturing
Lou Prout	Manufacturing Eng.
Charles Hilde	Packaging/Shipping
Ernest Ross	Laboratory
Linda Kazascky	Marketing
Tom Jurge	R&D
Curt Williams	Maintenance
Bill Andrews	Human Resources
Jim Jurgens	Technical Services

To: Distribution October 12, 1992

Re: 1992 Audit Schedule From: Jim Lamprecht

This tentative schedule is intended to give managers and supervisors advance notice of audits scheduled in their areas. You will be notified approximately ten to twelve days prior to the audit. The size of the audit team will depend on the scope and department being audited but will never exceed three auditors.

MONTH	AUDIT TYPE-SCOPE	DEPARTMENT
January	Document requirement	Training
January	Lab doc. control/procedures	Laboratory
February	Calibration procedure + training	Metrology
March	Procedure-change control plus work instructions.	Production Cntrl.
March	Traceability	Quality Assurance Production
April	Sample analysis procedures	Laboratory
April	Packaging/shipping procedures	Shipping
April	Documentation, vendor evaluation	Purchasing

Sample 2

To: Bruce Middleton
Human Resources

December 10, 1992

From: Jim Lamprecht

Re: Training documentation requirements as per paragraph 4.18 of ISO 9001.

I will be conducting an audit of your department on January 18, 1993. The audit will focus on the current documentation system. The reference for the audit will be paragraph 4.18 of ISO 9001 which reads as follows:

> **4.18 Training**
> The supplier shall establish and maintain procedures for identifying the training needs and provide for the training of all personnel performing activities affecting quality. Personnel performing specific assigned tasks shall be qualified on the basis of appropriate education, training, and/or experience, as required. Appropriate records of training shall be maintained (see 4.16). [Note: *Paragraph 4.16 should also be included for cross-reference. Some auditors simply include the whole ISO standard which, in the case of ISO 9001, consists of only seven pages.*]

I will forward an Audit Plan to you by January 10, 1993, along with a tentative schedule for the audit. The meeting-audit should last approximately 1-1.5 hours. If this schedule is not suitable, please feel free to contact me at X-5555.

Jim Lamprecht
Lead Internal Auditor

cc: Appropriate managers

Sample 3

The following pages are an example of how you might want to format your audit report findings.

To: Manager

From: Auditor or audit team

Re: Audit #92-05: Sampling Procedure Book

Purpose

The purpose of the audit is to review all of the Sampling Procedure Books as to completeness and up-to-date status (paragraph 4.5.1 and 4.5.2 (document control) of ISO 9001) of Inspection Procedures.

Scope

All 52 Sampling Procedure Books will be reviewed according to SP-A-1, III.E.1.

Procedure

All books found were audited by the same method. The Master Index date was checked; then five randomly selected SP's were checked as to the current revision letter.

Observations

- Sampling Procedure #18, 22 and 37 were not found.
- Twenty two percent of all Sampling Procedure Books had the wrong Master Index sheet.
- Twelve percent of the SPs had outdated revisions number.
- Thirty percent of the SPs had numerous uncontrolled handwritten notes.

Recommendations

1. Update SPs having handwritten notes and the Master Index sheet.

2. Your suggestion to computerize (word processing) the SP book should be implemented as soon as possible. It could certainly reduce your paper work. However, how will you inform marketing of all your updates? As you know, we are required to inform some of our clients of any sampling and testing updates.

Congratulations on the excellent job of maintaining the SP Books. The number of nonconformances has significantly been reduced from our last audit #91-11.

Sample 4

It is always a good idea to ask the people you audit to critique the audit process. The following sample page illustrates how you might want to evaluate the effectiveness of your audit. You might want to adapt or otherwise modify the questionnaire and questions to best suit your needs.

Quality Audit Program Evaluation

It would be helpful if you would add any constructive comments or observations concerning your experience as an auditee. You do not need to identify yourself by name. Please indicate your position.

____ Supervisor   ____ Department Head
____ Manager   ____ Director

Department ________________

Comments:

Please evaluate the following questions using the following five point scale: 1 = Strongly disagree, 2 = Disagree, 3 = No opinion, 4 = Agree, 5 = Strongly agree.

Questions	1	2	3	4	5
1. The audit was a fair appraisal of my operation.					
2. Interference with operations was at a minimum.					
3. The audit did not take long.					
4. I was notified prior to the audit and agreed to the time.					
5. The auditor was objective, fair and listened to my opinions.					
6. The auditor only reported discrepancies which were observed and confirmed by me.					
7. Disagreements were resolved equitably.					
8. The auditor was knowledgeable and qualified.					
9. The auditor(s) was/were prepared and proceeded in an organized and efficient manner.					
10. The auditor was considerate of our time constraints and was flexible.					
11. The audit was comprehensive and thorough.					
12. Minor deficiencies were clustered into a global finding.					
13. The auditor clearly distinguished factual (objective) evidence from hearsay.					
14. I was kept informed of the audit observations during the audit.					
15. I was the first to know the audit results in area of responsibility.					
16. Corrective actions will be implemented.					
17. The audit report did not embarrass me or make me feel defensive.					
18. The audit helped me to more effectively control quality and reduce costs.					
19. The audit helped me to better perform my job.					
20. I think the audit program is helpful and is a worthwhile effort.					
Average					

Sample 5

AUDIT FINDING RESPONSE FORM

Audit # 92-02 Finding # 1 Date: May 18, 1992

Audit Title: Document control/Change control as per ISO 9001 § 4.5.1 and 4.5.2.

Finding: Sampling procedures Lab 01-D-132 and Lab 12-F-13 were not up to date.

Cause of the problem:

Action(s) to be taken to correct and prevent recurrence:

By whom and when will corrective action be implemented?

Prepared by:______________________________ Date:________________

For internal lead auditor only

Follow-up audit required: No_____ Yes_____ By when ______________

Audit finding closed? Yes_____ No____

Reason:____

Sample 6

Findings Status Report

Definitions

Open findings	Findings reported on audits.
In-process findings	Findings that are not yet due or not yet closed.
Overdue findings	Findings which are more than 30 days overdue.

Open findings	11
In-process findings	09
Overdue findings	02
Total	22

Some auditors summarize their status report by department as follows:

Department	Open	In-process	Overdue
Purchasing	2		
Manufacturing	5	5	3
Laboratory	5	2	
Shipping	2		

Auditor: ______________________ Date: ______________________

Page 1 of 2

Findings Status

Finding # and date	Finding Summary	Status
List department and audit # and date	*Brief description of finding(s). Cite ISO paragraph(s) or relevant document whenever possible (i.e., quality manual, manufacturing third tier, etc.)*	*Cite status*
Heat treatment/ 91-11 11/12/91	No procedures for nonconformances ISO 9001/4.13.	Open
Manufacturing/ 92-11 11/18/92	No work instruction for corrective action ISO 9001/4.9 and 4.14.	Overdue

Auditor: ____________________ Date:__________________ Page 2 of 2

Conclusions

When conducted properly, internal quality audits do provide valuable information which *should* allow for evaluating the effectiveness of the system. Internal auditors should always remember that the audit process should be a two-way communication process during which information on how to improve the system can be gathered. Internal audits, or first-party audits should not be confused with third-party audits nor should they be necessarily conducted as third-party audits. Certainly, there is much to be gained by conducting internal audits using the procedural rigor of third-party auditing which is required by the ISO standard (see for example ISO 10011-1, 2 and 3 for further information). The next chapter will present various third-party audit scenarios. These scenarios are designed to educate the reader on how to handle third-party auditors as well as get a glimpse on what "to look for" during an audit.

8 A Look at Third-Party Audits

It has been my experience that in most cases, third-party audits tend to be very tense events; at least during the first couple of hours. They need not be of course. The mood of an audit will very much depend on the temperament of the auditors. Austere auditors who monotonously and rigidly follow their checklist, reciting one question after another, can bring on an unpleasant aura to the audit process. On the other hand, clients have also told me that their internal and external audit experiences were an absolute delight.

To anyone who is about to be audited, I offer the following suggestions:

- Know the standard inside out. This should be relatively easy to accomplish particularly if you are a member of the ISO 9000 implementation team.

- Be sure to order the ISO 10011-1, 2 and 3 series which outlines how to conduct and manage third-party audits as well as states various criteria for auditor competence and general traits. [Note: These standards can be ordered from the American Society of Quality Control in Milwaukee, Wisconsin.]

- Do not be intimidated by the auditor(s). Do not hesitate to question their particular interpretation of a specific clause. However, do try to be diplomatic and do not question *every* interpretation (hopefully this should never happen, particularly if you have done a good job documenting and referencing your system).

- If the auditor fails to reference the particular ISO paragraph(s) associated with an alleged nonconformance, remind him/her to please cite you the particular clause (s)he has in mind. Please note that an auditor does not always have to cross-reference a particular ISO clause with every nonconformance. *Nonconformances can be cited as a result of inconsistancies within the quality system.*

Do I Have to Answer That Question?

You should always attempt to answer all questions even if the answer is: "I don't know but I will find out who can answer your question." That is in fact a very good answer, for you should never guess or try to bluff your way out of a question. If you do not know the answer, simply say so and try to find out who can best answer the question.

In some instances you might be asked to answer a question which you feel is not pertinent to the standard. Examples of questions that might be considered peripheral to the standard would include:

1. "Show me how you rate your suppliers?"
2. "What is your grading scale for vendors?"
3. "How do you know two days later that your instrument is still calibrated?"
4. "Has this person been trained to do this particular job?"

The first two questions are variations of the same theme and are supposedly designed to assess whether or not the supplier has addressed paragraph 4.6.2 (Assessment of Sub-contractors). The clause however does not state that the supplier shall have a rating scheme. The questions are usually derived from the ISO 9004 Guidelines where recommendations are offered. If we refer to paragraph 9.3 of ISO 9004/ANSI/ASQC Q 94, one can read that:

> "The methods of establishing this capability (supplier capability that is,) *may* include any combination of the following:
>
> a) on-site assessment and evaluation of supplier's capability and/or quality system;
> b) evaluation of product samples;

c) past history with similar supplies;
d) test results of similar supplies;
e) published experience of other users."[1]

Obviously, in this case, the auditor is influence by paragraph a) which suggests "on-site assessment." You may evaluate your suppliers differently, and in fact you are likely to rely on c) "past history." Consequently, if your quality manual states that suppliers are evaluated using various schemes including past history (or perhaps only past history), you only need to refer the auditor to the appropriate paragraph in your quality manual.

Question 3 is a legitimate question if a bit convoluted. It appears to be designed to verify if the supplier checks adequacy of calibration frequency or perhaps sub-paragraph g) of 4.11 (assessment of previous inspection and test results when instrument is found to be out of calibration). If a gauge or instrument is that unstable, your are likely to recalibrate it prior to every measurement and should say so in your procedures.[2] As for paragraph g) of 4.11, which attempts to address the situation when an instrument falls out of calibration while within the calibration interval, that is more difficult to answer except perhaps to recall all parts (which is very expensive), or to warn the customer of possible problems.

The last question could be a legitimate question (but poorly phrased nonetheless). It all depends on what type of job the auditor is refering to. If you can, or have, demonstrate(d) that the job in question does not affect or relate to quality, you should have no difficulty defending your position. If, however, the job in question affects a lab assistant who is performing quality-related chromotography analysis, you will

---

[1] ANSI/ASQC Q 94, p. 10.

[2] Laboratory technicians often routinely and perhaps unnecessarily, recalibrate some instruments prior to every measurement. Some equipment are self-calibrating. Be sure to explain these procedures in your level two and/or three documents.

have a much more difficult task unless, of course, the assistant in question happens to be an analytical chemist.

How to Formulate Questions

Although the above four questions could be considered adequate, they nonetheless indicate that the auditor asking the questions might have had little experience auditing quality systems. Experienced auditors tend to use different techniques when trying to find out if a particualr procedure is indeed implemented. My favored technique is to use what I have come to refer to as the "funnel approach." The purpose of the "funnel approach" is to first ask a general or broad question relating to a particular theme. The auditor should next follow up on his question by asking a more specific question, which should be derived from the contents of the auditee's first answer. For example, an auditor witnessing an operator checking an I/C board could ask the following two questions *to the operator*:

> "What does the verification activity entail?"
> (*The answer could include: "I follow instructions written on the schematics," or "I go to the inspection sheet," or a specific explanation of how the board is inspected. Each answer can lead to a different type of question. What follow-up question could you ask to the first answer ("I follow instruction . . . "); to the second answer?*

If your main objective is to verify training activities, you can then follow-up with the following question:

> "How do you know what to verify for each board?"

If you do not hear any reference to training, you might have to go to the next question and ask:

"Has anyone explained to you or trained you on how to verify boards?"

Of course, the auditor could have asked that last question first and skip the other two but then he would have missed a lot of information. The principle is rather simple. An auditor can mechanically go through a list of so many questions and record a "Yes," "No" or "N/A" answer and move on to the next question; or he can try to learn how the system is really working.

To help you better prepare for a third-party audit, I am including the following two case studies.

Case Study/Exercise 1 (Allow 45-60 Minutes)

This first case involves a chemical plant which produces one product that is sold to half a dozen clients. For the most part, the product is delivered directly to the customers via pipelines. Approximately half-a-dozen raw materials are delivered to the plant either via trucks, railroad cars or pipelines. The raw materials are stored in several storage tanks and processed (blended, mixed, heated, etc.) via a splitter unit which is *controlled* by a Honeywell control process system.[1] The company in question wants to achieve ISO 9002 registration.

During one of the pre-assessment visits, the auditor is taken to the Honeywell control room which, (s)he is told, "is the heart of our operation." Upon entering the large room the auditor notices a large

---

[1] The use of computers in the process industry, particularly the chemical industry, goes back to the early 1950s. David E. Noble explains that "By the 1950s the first analog-computer-controlled industrial operations appeared in the electrical power and petroleum refinery industries. Computers were used to monitor performance, log data, and instruct operators." It is interesting to note that in the 1920s, the dairy industry, which was one of the first industries to go into continuous process production, "had developed ways of carefully controlling the temperature of its product in process, to comply with the federal pasteurization laws introduced in the 1920s." See Noble, *Forces of Production* (New York, Oxford University Press, 1986), pp. 59-63.

computer console surrounded by a couple of rows of what appear to be monitoring equipment. Next to the console, one can see a few manuals labelled "Honeywell." Three individuals surround the console and are chatting with a man sitting in front of three computer screens. As the auditor is about to be introduced, (s)he notices that the "computer operator" is staring at the screens and occasionally touching one of the screens.

The "computer operator" is introduced as the process control operator. The other three individuals are introduced as being from maintenance (two individuals) and from engineering (a process engineer). Before being asked any question, the process control operator proceeds to explain what he does.

Listening attentively, the auditor learns that the whole process chemistry is modeled and controlled from the three screens which represent schematics of the splitter unit. Each screen consists of several windows which can be accessed by simply touching various portions of the screen. Several key parameters, such as temperature, pressure, flow rate and pH, can be monitored and accessed from any of the screens. Temperature, for example, can be checked throughout the process at various critical points. Product samples are taken at various locations by operators and submitted to the laboratory for analyses. Samples are also analyzed by on-line analyzers.

The auditor learns that the whole splitter unit is sensitive to rapid and extreme meteorological changes (characteristic of the region) which obviously affect ambient temperature. As far as the final customer is concerned, the only important characteristic is that the product be at least 99.5 percent by weight.

The interview continues for approximately forty to forty-five minutes during which time much more information is gathered. From the information contained in the above paragraphs try to address the following questions:

1. What questions relating to ISO could you ask? As you formulate your questions, which ISO paragraph(s) would apply?

2. To whom should you ask your questions?

3. What other department(s)/functions might you want to interview?

4. What would you consider to be a nonconformance? Naturally that is difficult to answer since you do not know what the answers will be nor do you have a description of the quality assurance system in place. Nonetheless, attempt to analyze what you would consider to be an acceptable answer. Do not attempt to go beyond the standard's requirements. You might find this exercise more valuable if two or more people work on it independently and then compare notes.

Case Study/Exercise 2 (Allow 45-50 Minutes)

This case study involves a manufacturer of precision bearings. The company would like to achieve ISO 9001 registration. A quality manual written to address ISO 9001 had just been printed and circulated to members of the ISO 9000 committee. After visiting various departments, the auditor is taken to an area where a particular type of bearing is manufactured. On the way to the area in question, the auditor notices several plastic containers along a wall. Each cart is full of parts and has documentation which appears to be a traveller. Curious, the auditor asks his host: "What is this?" pointing to the parts. "I don't know, let's find out." Upon inspection, the auditor learns that the document in question is indeed a work order/instruction sheet. The work order sheet (which consists of several pages), indicates, among other things, where the part was last processed, what was the nature of the job and where the part is to next be processed. All steps and processes are numbered and coded from the first to the last process/step. Having been assured by his host that all parts regimentally follow each step sequentially—and thus cannot possibly "skip" a process—the auditor asks to see if indeed the parts in question had just "arrived" from the previous process as specified on the work instruction. Upon inspection, and after pausing for a brief moment, the host sheepishly admitted that, in this particular instance —a rare occasion indeed—the parts had "skipped" a couple of processes only because a particular machine was momentarily down for unscheduled maintenance. When the machine comes up again, the parts will be routed to ensure that the two processes in question are not missed.

Wanting to know a little bit more, the auditor asks if he could talk to one of the operators. Assured that he could, the auditor goes to the nearest operator and asked her if she could explain the work order in question. The following (recreated and abridged) dialogue ensued:

Auditor: "Good afternoon, I am John Lamp. I have been asked to conduct a pre-assessment audit for your company. I was noticing these parts (pointing to the wall). Could you explain to me what is to be done to them?"

Operator: "Oh, they will be picked up to go to grinding."

Auditor: "How do you know?"

Operator: "Its written on the work instruction. See . . ." The operator proceeded to explain how the work order/instruction guided parts throughout the plant.

Auditor (Pointing to the "missing" two steps on the work order): "But what about these two steps, I don't see an inspector stamp. Can parts skip processes and be re-processed later?"

Operator: "Oh yes, that happens quite often. Sometimes machines are down and we can't wait to process the part, so we push them through and reprocess them later."

Auditor: "Is that okay? Can the parts follow processes in sequences different from those specified on the work instruction?"

Operator: "It depends. For some parts we can't really skip sequence but sometimes we have to, you know . . . For some parts it does not matter in which order you do them."

Auditor: "Who decides what should and should not be done."

Operator: "The foreman."

Auditor: "Thank you very much, we have to go to the Numerical Control (N/C) section."

Operator: "No problem."

Arriving at the N/C section the auditor is introduced to one operator. The host asks the operator to explain her job. The N/C operation controls a drill. As the operator begins to explain her job she grabs a set of computer-generated N/C instructions which are about three pages long. The instructions are full of handwritten notes, comments and command changes. When asked who inserted the changes, the operator explains that she wrote some and someone else wrote most of them. She then proceeds to explain that the handwritten additions/insertions were only temporary and that they would be programmed in the machine at a later date. When asked if a new printout would be generated at a later date, she answers "yes" but can not find the one which applies to the current run.

The operator then explains that she had to keep a close eye on the drill because it is important NOT to use a drill that was "past its life."

Auditor: "How do you know when to change the drill?"

Operator: "I know by looking at the parts when it's time to change the tool."

Auditor: "Is this machine under preventive maintenance?"

Operator: "No. They come sometimes when things go wrong. Its a pretty good machine, few things go wrong."

---

1. What ISO paragraph would/could you apply to the above scenario?

2. What other questions could you ask?

3. Are there any (possible) nonconformances? If so, where and what ISO paragraphs would you cite?

Suggested answers for the both scenarios are provided in Appendix E.

Having briefly reviewed the audit process one must next ask how reliable and repeatable the audit process is. By *reliability* of the audit process I mean how *confident* one can be that the results of an audit are accurate. If an audit is reliable, will it necessarily be repeatable? In other words, will the same—or nearly the same—results be derived by another registrar. One would hope so. These are difficult questions which I will *attempt* to answer. I have emphasized *attempt*, because the only way to test for reliability and repeatability would be to somehow convince all registrars to participate in a study; something they are not likely to agree to in the near future.

The Audit Process—How Reliable and Repeatable?

Whenever a registrar audits a company, both the registrar and the company are exposed to the standard statistical risks known as the *Producer's and Consumer's risks* (also known as *alpha and beta risks).* Indeed, when a registrar audits an organization, it cannot possibly hope to interview everyone, and review, read or otherwise verify every activity. To do so would be prohibitively expensive and of dubious value. Consequently, any auditor must rely on sampling. Readers familiar with acceptance sampling or basic statistics will quickly recognize that whenever samples are taken, risks surround the estimation of statistical parameters. Thus, without going into a detailed statistical analysis, it can be stated that whenever a registrar audits a company, a working hypothesis (known as a null hypothesis, H_o) could be formulated. One of the many null hypotheses that could be stated could read as follows:

H_o : "Company A will pass the audit (and thus achieve certification)."

The alternative hypothesis (H_a) state that the company would not pass the audit. In order to verify these hypotheses, auditors sample the quality assurance system of the said company and arrive at a conclusion. But because auditors can only sample the quality system, both the company and the registrar (and indirectly customers to the company) are exposed to risks whenever the above hypotheses are "tested." The company's risk (or probability) of having its quality assurance system rejected when in fact it should be accepted is known as the *Producer's risk*. Similarly, the registrar runs the risk (probability) of accepting the quality assurance system when in fact it should be rejected. This probability of *mistakenly* registering a company to one of the ISO quality assurance standards when in fact the quality system should be *rejected*, is bound to affect the company's customers (and the registrar's reputation) and is therefore known as the *Consumer's risk*.

Can these risks be assessed? Probably yes, if enough time and data are collected. In fact, I believe that some registrars do in fact publish probability of acceptance based on their document sample size. I do not, however, have enough information to determine whether or not the statistics are worth the effort or are of interest to anyone except perhaps to those interested in esoteric studies.

Given the subjective nature of the ISO 9000 series of standards, one should not be surprised to learn that risk and variability are present whenever audits are conducted.

Sources of Variability

There are at least three major sources of variability which might influence any audit process:

> 1. *Psychological profile of the auditor*
> 2. *Training and experience auditing quality systems (as opposed to product audits)*
> 3. *Industry experience that is likely to influence an auditor's point of view*

Another variable which is likely to affect the audit process is number of audits conducted per month. An auditor who conducts too many audits per month (say, more than 10-12 days per month), is likely to suffer (sooner or later) burn-out which *will* influence his/her ability to listen and thus conduct effective audits. Let us examine each of the above sources of variation.

Auditor's Psychological Types

Psychologists generally recognize four basic psychological types known as: Intuitive Feeling types, Intuitive Thinking types, Sensible Judging types, and Sensible Perceiving types.[1] These basic types can be further broken down into sixteen categories summarized in Table 8.1. Table 8.2, duplicated from *Life Types* by S. Hirch and J. Kummerow, provides amateur psychologists with an ingenious one-line description of the meaning behind the various acronyms.

---

[1] For further details on the *Myers-Briggs Type Indicator* see David Keirsey and Marilyn Bates, *Please Understand Me* (Del Mar: California, Prometheus Nemesis Book, 1984) and David Keirsey, *Portraits of Temperament* (Del Mar: California, Prometheus Nemesis Book, 1991). Numerous other books are also available in most book stores.

Anyone interested in knowing what their psychological profile might be can either subject themselves to the Myers-Briggs test or conduct a quick self-assessment test provided in most temperament types books (such as David Kernsey's for example.)

Table 8.1 Temperament Types

	Sensing	*Sensing*	*Intuitive*	*Intuitive*	
Judging	***ISTJ***	***ISFJ***	***INFJ***	***INTJ***	*Introvert*
Perceiving	***ISTP***	***ISFP***	***INFP***	***INTP***	*Introvert*
Perceiving	***ESTP***	***ESFP***	***ENFP***	***ENTP***	*Extravert*
Judging	***ESTJ***	***ESFJ***	***ENFJ***	***ENTJ***	*Extravert*
	Thinking	*Feeling*	*Feeling*	*Thinking*	

Table 8.2 Life Types

ISTJ I Save Things Judiciously	**ISFJ** I Serve Family Joyfully	**INFJ** Inner Nuances Foster Journeys	**INTJ** It's Not Thoroughly Justified
ISTP I See The Problem	**ISFP** I Seek Fun and Pleasure	**INFP** I Never Find Perfection	**INTP** It's Not Theoretically Possible
ESTP Everyone Seems Too Proper	**ESFP** Extra Special Friendly Person	**ENFP** Every day New Fantastic Possibilities	**ENTP** Each New Thought Propels
ESTJ Execution Saves The Job	**ESFJ** Extra Special Friendly Joiner	**ENFJ** Everyone Needs Fulfillment and Joy	**ENTJ** Executives Need Tough Jobs

Source: Hirsch, S and Kummerow, J. *Life Types. (New York: Warner Books, 1989)*

A cursory look at Table 8.2 reveals that an ISTP (I See The Problem), or INFP (I Never Find Perfection) auditor is not likely to audit a quality assurance system from the same perspective as an ENFJ or ESFP auditor. Such a diversity of opinions is hopefully minimized or filtered out via external and internal training.

Auditor's Training/Experience

All ISO auditors are supposed to have been through a thirty-six hour approved training course. These courses are designed to teach or retrain auditors on the art of auditing ISO type quality assurance systems. Having passed a written test, auditors must next conduct three to five audits before being recognized as auditors.[1] Most auditors in the U.S are either registered by the Institute of Quality Assurance in the U.K., the Registrar Accreditation Board of Milwaukee, Wisconsin (or both), or the Canadian board. Other accreditation agencies would include the French Association for Quality Assurance (AFAQ). Each European country has or will soon have its own accreditation agency.

In some cases, where the audit might also include product certification, the auditor also has to list his or her expertise in particular industries. Most accreditation boards require the applicant to list his area of expertise. Knowledge of a particular industry is likely to influence an auditor's perception of the quality assurance system. However, in cases where the the documented quality assurance system is audited, an auditor's experience in a particular industry should not affect the results.

The subjectivity of any audit is therefore a function of at least three variables: psychological temperament of the auditor, training and auditing experience and knowledge of the industry (See Figure 8.1). If in addition, one wants to measure audit reproducibility, variables

[1] Be aware that some registrars seem to train their auditors during pre-assesments. This is perfectly legal; however, you should ensure that you are not billed for this training session.

such as the registrar's internal training program, the registrar's philosophical interpretation of the standard and other intangible variables would have to be included.

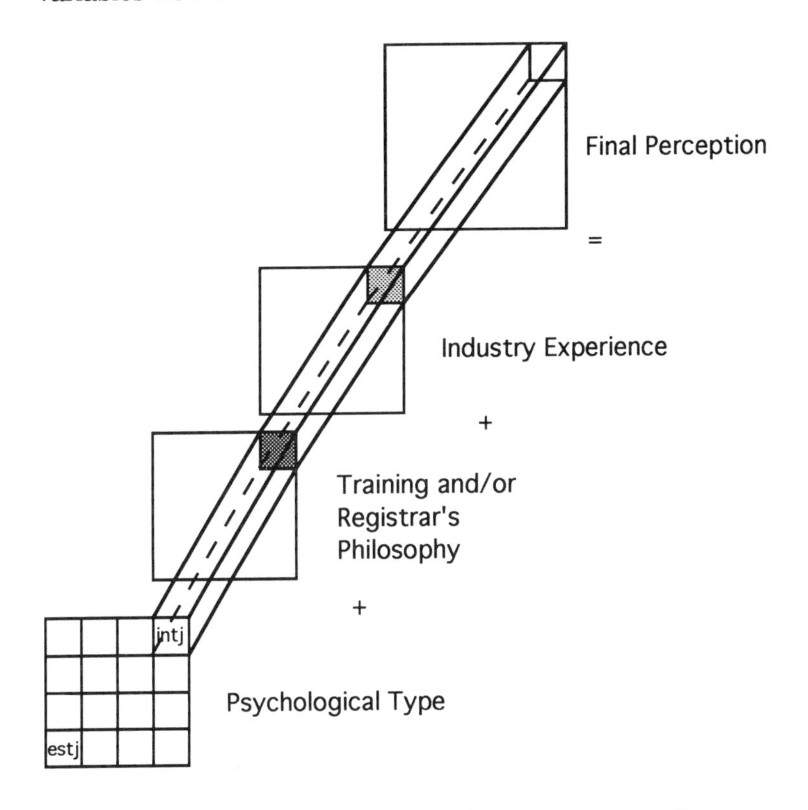

Figure 8.1 Factors Influencing an Audit

On Variability and the Need for Standardization

The need and importance to standardize and formalize audit procedures became apparent to me when I was given the opportunity to conduct an audit training seminar in France in early 1990 for a U.S. multinational corporation. The project consisted of training approximately sixty to seventy auditors using a Supplier Quality System Survey developed in the U.S.

The survey consisted of a seven-part questionnaire (A-G). Each section in turn contained several questions ranging from as few as six questions for section F (Calibration & Measurement Control) to as many as sixteen questions for section A (Quality Management). Each question was also assigned a weight of 5, 8, 10, 12 or 15, directly proportional to its perceived importance (no one knew who had assigned the weights or why certain values were assigned to particular questions). Finally, the auditor was to assign a rating ranging from 0.0, 0.3, 0.5, 0.6, 0.8 or 1.0 depending on whether or not the supplier addressed a particular item and how well. A zero meant that the question was not addressed whereas a 1.0 meant that the supplier was "world class", a somewhat difficult word/concept to translate into French.

The program called for a series of two-day seminar/workshops for all sixty-five participants on how to interpret and grade the survey as well as how to conduct audits (question formulation, listening, etc.). Some of the participants were then grouped into half a dozen teams of three to four members, including the author. Each team was to then visit a supplier and practice the skills they had learned during the seminar.

During the many visits to various suppliers throughout France, it became evident that although all participants had received the same training, scores would vary greatly—by as much as three points in some cases which could lead to very large deviation when the weighting factor was multiplied. It soon became evident that before recording our final scores each section would be reviewed and a consensus final score would be reached. Although the process was tedious, the discussion process was most beneficial to all. It was during that time that I decided to conduct an informal quantitative analysis of the results.

Although I have not performed a formal library search on the topic, to my knowledge, no quantitative studies designed to assess variability among auditors have yet been published. The results presented below are extracted from an as yet unpublished study conducted by the author during one supplier audit in 1991. The audit team consisted of three auditors: one auditor had several years experience in manufacturing, the other had no experience in auditing but had consulting experience and was familiar with the principles of quality assurance. The author was the third auditor. The questionnaire was the same seven-part questionnaire described above. The audit was conducted over a two-day period after which time all scores were compared, and a consensus score was reached (not included here). Raw scores and a chart for section B, which consists of eight questions, are shown in Figure 8.2. The associated frequency diagram is printed to the side. The sample size, *n*, is three, since there were three auditors. The number of samples taken, k, is eight (eight questions). Although the customary upper and lower control limits could be computed, they are not included here simply because their significance is of dubious value. One should note that an Analysis of Means would probably be more meaningful than the usual Shewhart chart.[1] As can be seen from the graph, the audit process for section B exhibits substantial within-auditor variability (particularly for question 3) see also Figure 8.3 where the same information is represented using Box-Whisker diagrams.[2] Figure 8.4 summarized the auditors' scoring across section A and Figure 8.5 represents scores for sections B through G. Notice the substantial auditor variability within questions and across sections. The dotted line drawn at 0.66 represents the minimum scores required for each section to achieve approved supplier status. Obviously, even after accounting for within auditor variability, this particular supplier would still not make the "approved" ranking!

---

[1] For a good discussion on ANOM (Analysis of Means), see Ellis R. Ott, *Process Quality Control* (New York: McGraw-Hill Book Company, 1975) and Donald J. Wheeler, *Understanding Industrial Experimentation* (Knoxville, TN: Statistical Process Controls, Inc., 1988).

[2] Auditor variability within section and across questions within sections can also be graphically demonstrated with the use of Box-Whisker diagrams.

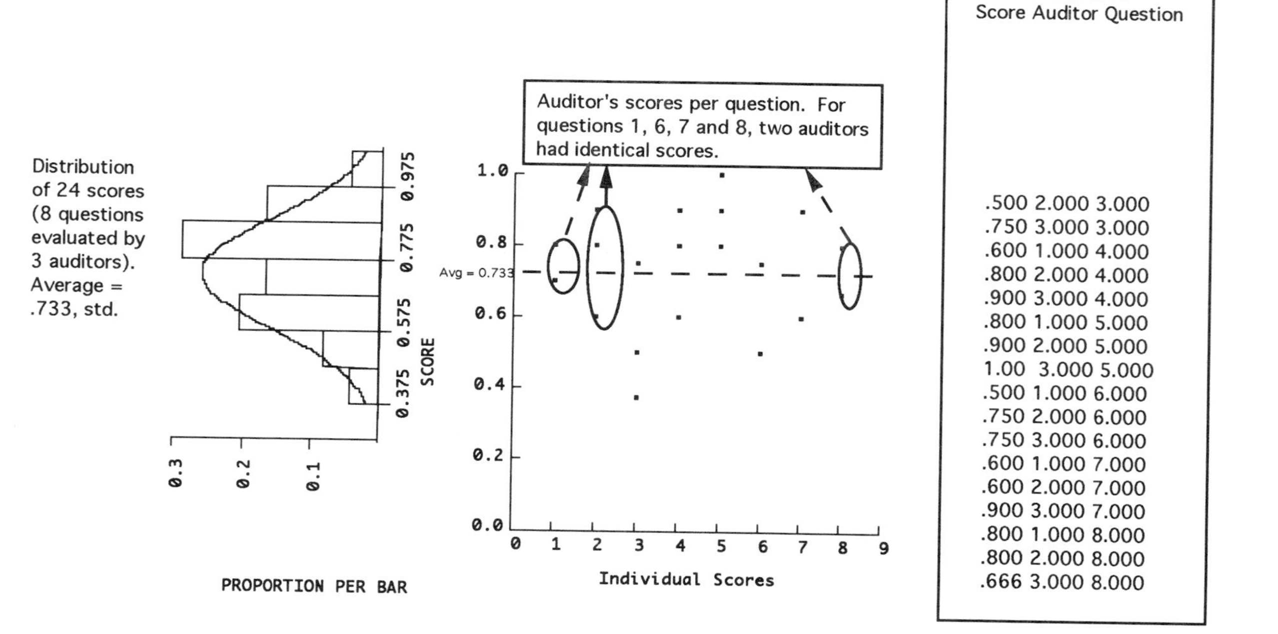

Raw Data

Score	Auditor	Question
.500	2.000	3.000
.750	3.000	3.000
.600	1.000	4.000
.800	2.000	4.000
.900	3.000	4.000
.800	1.000	5.000
.900	2.000	5.000
1.00	3.000	5.000
.500	1.000	6.000
.750	2.000	6.000
.750	3.000	6.000
.600	1.000	7.000
.600	2.000	7.000
.900	3.000	7.000
.800	1.000	8.000
.800	2.000	8.000
.666	3.000	8.000

Figure 8.2 Scores for Section B (Eight Questions)

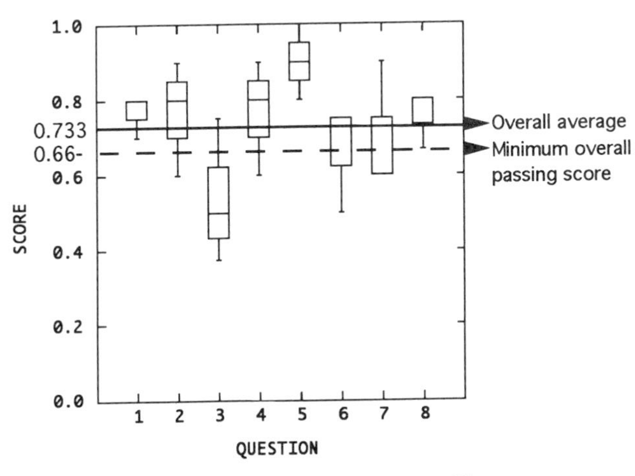

Note: The boxes are known as Box-Whisker diagrams. They represent the amount of "auditor variation" per question. For example, auditors generally agreed on question 1 (narrow box) but disagreed on question 3 (wider box).

Figure 8.3 Box and Whisker Diagrams for Section B

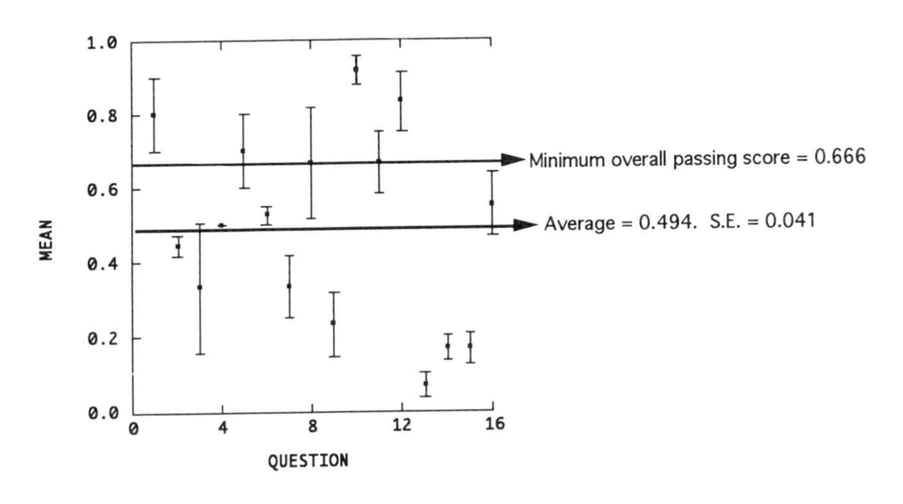

Figure 8.4 Scores for Section A (16 Questions)

Figure 8.5 (next page) Scores for Sections B-G

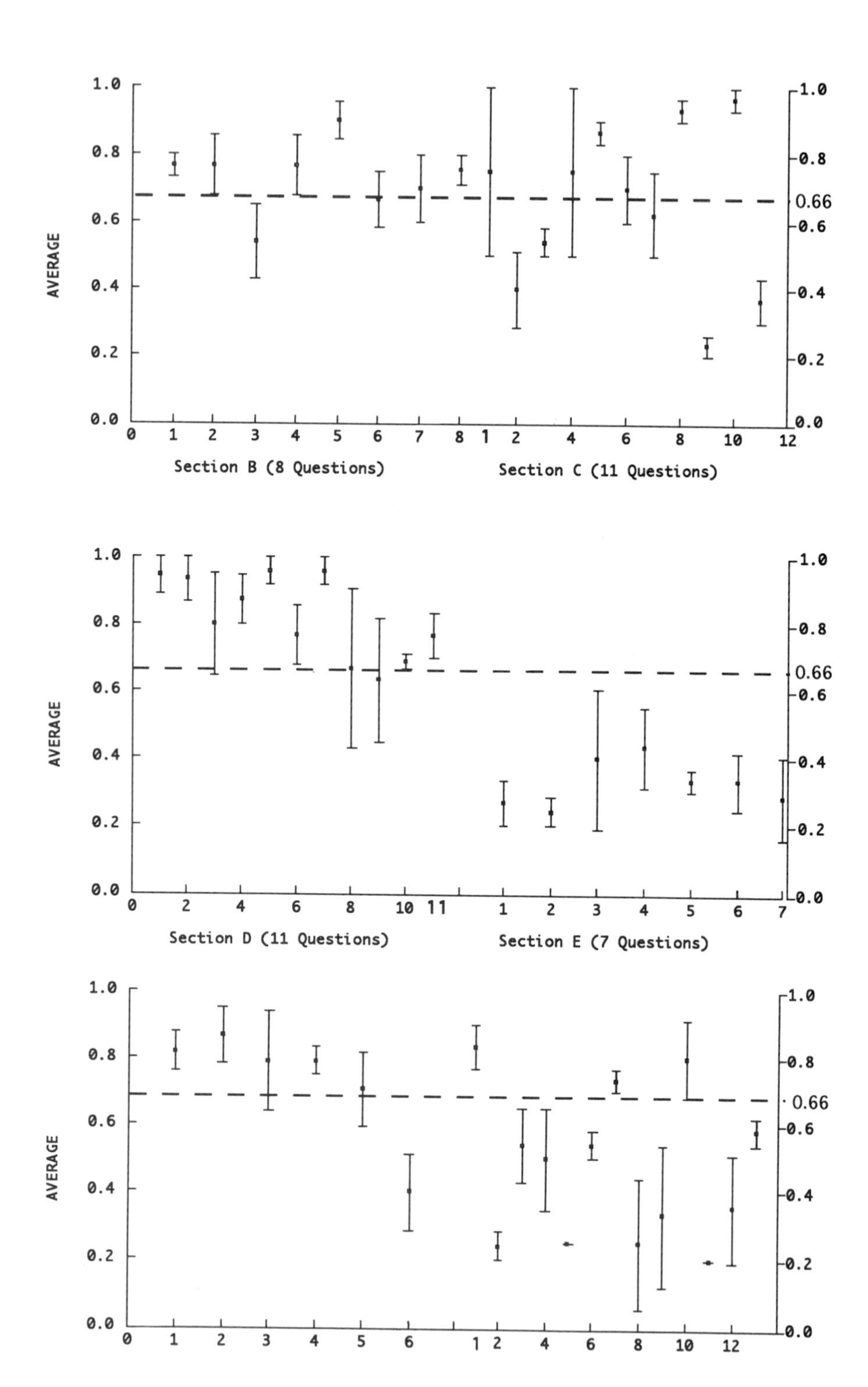
AVERAGE
Section B (8 Questions)
Section C (11 Questions)
0.66
AVERAGE
Section D (11 Questions)
Section E (7 Questions)
0.66
AVERAGE
0.66

Most Often Made Mistakes When Audited

The following list is an abridged and modified account taken from a flyer found in the office of a certified public accountant and entitled "The 10 Biggest Mistakes You Can Make at an Audit" (anonymous and undated). Although the flyer was referring to IRS audits, similar principles apply to third-party ISO 9000 audits. Nevertheless, the suggestions listed below are only meant to be guidelines, indeed, third-party audits should *never* be conducted as IRS audits. There certainly are similarities between the two types of audits: in both cases the auditors (IRS or ISO) are trying to determine whether your submitted documented evidence matches reality (facts). However, whereas the IRS audit is likely to result in a punitive monetary fine, the ISO audit will, in most cases, allow for—in fact require—that corrective actions be implemented prior to issuing a certificate. In addition, whereas you are summoned by the IRS, you purchase the service of a third-party registrar. Finally, the anxiety and paranoia usually associated with IRS audits should not be experienced during ISO audits. Third-party audits are intended to be helpful exercises—their objective is to, sooner or later, offer registration.

1. Not getting professional help. This can be a time consuming and, hence, costly mistake. Indeed, casual interpretation of the often vague and all encompassing verbiage found in the ISO 9001, 9002 or 9003 (ANSI/ASQC Q9001, Q9002 or Q9003) standards can lead to some misinterpretations resulting in lengthy and often unnecessary (and thus costly) procedure writings.

2. Being too responsive to an ISO assessor's questions. A good rule to follow with most third-party auditors, particularly if they strictly and solely adhere to a checklist-type audit, is to **avoid volunteering information**. An answer that goes far beyond the specific question could eventually lead to other questions that would never have been asked in the first place. **The right way to field questions** is to keep your answers brief and to the point. Be polite and courteous but never expansive.

3. Failing to prepare for the audit. The danger in going into an audit unprepared and disorganized is that the auditor might broaden the examination to include items not included in the audit notification letter. Avoid this by doing everything you can to expedite the audit process. Understand the issues involved. Try to anticipate questions. One of the best ways to establish credibility is to be knowledgeable, prepared and organized.

4. Including superfluous information. This generally will encourage an auditor to dig into other areas.

5. Not being truthful. If the auditor catches you being less than completely honest, your credibility will be damaged for the rest of the audit and the auditor will insist that you prove everything. It is always better to say "I don't know" or "'I'll get the information."

6. Not knowing your appeal rights. It is important to know that, if you should have a fundamental disagreement with an auditor, you can always appeal his/her decision.

Conclusions

Even if the ideal characteristics of an auditor could be cloned so as to produce the perfect audit team which would nearly eliminate disagreements and thus considerably reduce "auditor variability," subjectivity would not be eliminated simply because, in many cases, the instrument (i.e., the questionnaire), is either inadequate, poorly designed or inappropriate. In has been my experience that all too often questionnaires are quickly written and not tested for *validity* or *appropriateness*. Contrary to what might generally be believed, questionnaires designed for one industry or one application cannot be simply duplicated for another application or industry. Even though it is true that in many cases the same quality assurance principles may apply, questionnaires should be carefully planned to suit a particular need. One should ensure that each question is well stated and not subject to any misinterpretation. In some cases, questions may not apply or may even be preposterous within a certain context. And yet, questionnaires or checklists used as a substitute for interviews or verification activities are sometimes used as the most expeditious means to (unfairly) assess suppliers. If it is to be effective and well-respected, the audit process must be a well-thought-out process.

Having reviewed the audit process, we next present a brief preview of where the ISO 9000 series is likely to be heading in the next few years.

9 A Look at the Draft Updates to the ISO 9000 Series

The ISO 9000 series is to be reviewed every five years, however, because of delays the 1992 revisions will probably not be published until late 1993. The working groups in charge of updating the ISO 9000 document (SC2/WG 10), ISO 9001 (SC2/WG 11) and ISO 9002 and 9003 (SC2/WG 12) have done an excellent job clarifying the notoriously murky prose of the current (1987) edition. It is the intent of this chapter to offer to the reader a preview of the **Committee Draft** for the ISO 9000 series. The focus will be on the ISO 9001 standard (WG 11). **It is important to re-emphasize that the present review is of committee draft (CD) documents which, if approved, will then be circulated and reviewed as Draft International Standards (DIS). These DIS documents will then have to be approved prior to being released as official ISO documents.**

The ISO 9001 Draft Updates

Table 9.1 reveals that, although the structure and format has remained the same, some sub-paragraphs have been added and others combined. In the author's opinion, these additions do enhance the standard's readability and do help clarify what used to be nebulous phraseology. A typical enhancement can be found under the heading **Scope**. Rather than stating that "The requirements specified in this Standard are aimed primarily at preventing nonconformity at all stages from design to servicing," (ANSI/ASQC Q-91: 1987), the proposed change now reads as follows:

> "The requirements specified in this International Standard are aimed primarily at achieving customer satisfaction by preventing nonconformity at all stages from design through servicing."[1]

---

[1] Committee Draft ISO/CD 9001, p. 4.

Most changes are minor but they do help improve the standards readability. The authors have deliberately changed every reference to "procedure" to read as "documented procedures" thereby increasing

Table 9.1 ISO 9001 Draft Updates

ISO 9001 (1987)	ISO 9001 Update	Major Changes
0.0 Introduction	0.0 Introduction	
1.0 Scope and Field of Application 1.1 Scope 1.2 Field of Application	1. Scope	Reference to "achieving customer satisfaction." Field of application merged with scope.
2.0 References	2. Normative References	Stresses the need to use the most recent edition. Also refers to auditing guidelines ISO 10011 and measuring equipment ISO 10012.
3.0 Definitions	3. Definitions	Refers to ISO 8402 and expands definition of product to include hardware, processed material, software or combination thereof.
4.0 Quality System Requirements	4.0 Quality System Requirements	
4.1 Management Responsibility 4.1.1 Quality Policy	4.1 Management Responsibility 4.1.1 Quality Policy	Adds that "The policy shall be relevant to the supplier's organizational goals and the expectations and needs of its customers."
4.1.2 Organization 4.1.2.1 Responsibility and Authority	4.1.2 Organization 4.1.2.1 Responsibility and Authority	Emphasizes "process and quality system problems," rather than just quality problems.
4.1.2.2 Verification Resources and Personnel	4.1.2.2 Resources	Basically combines two (1987) paragraphs into one shorter paragraph. Drops specifications as to what constitutes verification activities.
4.1.2.3 Management Representative	4.1.2.3 Management Representative	Specifies that executive management with overall responsibility for quality "shall appoint a member of its management..." More specific as to reporting of the performance of the quality system.
4.1.3 Management Review	4.1.3 Management Review	Basically the same.

4.2 Quality System	4.2 Quality System 4.2.1 General 4.2.2 Quality System Procedures 4.2.3 Quality Planning	Notes in 1987 edition are now included with more details in the revised edition. More specific on quality plan.
4.3 Contract Review	4.3 Contract Review	Refers to "documents procedures" and clarified by stating "Each accepted invitation to tender, contract and order. . shall be reviewed . . ."
4.4 Design Control 4.4.1 General 4.4.2 Design and Development Planning 4.4.2.1 Activity Assignment 4.4.2.1 Organizational and Technical Interfaces 4.4.3 Design Input 4.4.4 Design Output 4.4.5 Design Verification 4.4.6 Design Changes	4.4 Design Control 4.4.1 General 4.4.2 Design and Development Planning 4.4.3 Organizational and Technical Interfaces 4.4.4 Design Input 4.4.5 *Design Review* 4.4.6 Design Output 4.4.7 Design Verification *and Validation* 4.4.8 Design Changes	Better structured and overall better text. Design output (4.4.6 c) is rephrased to state "identify those characteristics of the design that are crucial to the safe and proper functioning of the product such as operating, storage, handling, maintenance and disposal requirements;"
4.5 Document Control 4.5.1 Document Approval and Issue 4.5.2 Document Changes/Modification	4.5 Document *and Data Control* 4.5.1 General 4.5.2 Document Approval and Issue 4.5.3 Document Changes/Modifications	Does emphasize data. Basically the same except that "invalid and/or obsolete documents" shall be promptly removed.
4.6 Purchasing 4.6.1 General 4.6.2 Assessment of Sub-contractors 4.6.3 Purchasing Data 4.6.4 Verification or Purchased Product	4.6 Purchasing 4.6.1 General 4.6.2 Evaluation of Sub-contractors 4.6.3 Purchasing Data 4.6.4 Verification of Purchased Product a) Supplier Verification at Sub-contractors. b) Customer Verification of sub-contracted product	Clearer explanation of what is meant. Basically same intent except better organized.
4.7 Purchaser Supplied Product	4.7 *Customer* Supplied Product	Replaces confusing purchaser term with "customer."
4.8 Product Identification and Traceability	4.8 Product Identification and Traceability	None
4.9 Process Control 4.9.1 General 4.9.2 Special Process	4.9 Process Control	Special processes are now incorporated in 4.9. Maintenance of equipment is also included (e) in this paragraph

4.10 Inspection and Testing 4.10.1 Receiving Inspection and Testing 4.10.2 In-Process Inspection and Testing 4.10.3 Final Inspection and Testing 4.10.4 Inspection and Test Records	4.10 Inspection and Testing 4.10.1 General 4.10.2 Receiving Inspection and Testing 4.10.3 In-Process Inspection and Testing 4.10.4 Final Inspection and Testing 4.10.5 Inspection and Test Records	4.10.3 consists of only two sub-paragraphs a) and b) instead of a-d. No major changes
4.11 Inspection, Measuring, and Test Equipment	4.11 Control of Inspection, Measuring and Test Equipment 4.11.1 General 4.11.2 Control Procedures 4.11.3 Test Hardware and Software Checks.	Slight reorganization which enhances readability. Reference to ISO 10012. Minor rephrasing.
4.12 Inspection and Test Status	4.12 Inspection and Test Status	More condensed paragraph but same intent.
4.13 Control of Non-conforming Product 4.13.1 Nonconformity Review and Disposition	4.13 Control of Nonconforming Product 4.13.1 General 4.13.2 Nonconforming Product Review and Disposition.	None except for the reference to "documented procedures" instead of "procedures."
4.14 Corrective Action	4.14 Corrective *and Preventive Action* 4.14.1 General 4.14.2 Corrective Action 4.13.3 Preventive Action	Seems to be saying the same but with more words!
4.15 Handling, Storage, Packaging, and Delivery 4.15.1 General 4.15.2 Handling 4.15.3 Storage 4.15.4 Packaging 4.15.5 Delivery	4.15 Handling, Storage, Packaging, and Delivery 4.15.1 General 4.15.2 Handling' 4.15.3 Storage 4.15.4 Packaging 4.15.5 Preservation 4.15.6 Delivery	Basically the same except that the reference to preservation and segregation in 4.15.4 (1987) is now specified in 4.15.5 of the update. Not clear as to advantage of doing so.
4.16 Quality Records	4.16 *Control of* Quality Records	Slight rephrasing. Also recognize the use of electronic media.
4.17 Internal Quality Audits	4.17 Internal Quality Audits	None except reference to ISO 10011 in note.
4.18 Training	4.18 Training	Reference to "documented procedures" instead of "procedures."
4.19 Servicing	4.19 Servicing	Clarifies by specifying "Where after sales service is a specified requirement,.."

4.20 Statistical Techniques	4.20 Statistical Techniques	Attempts to improve paragraph by stating that "The supplier shall identify the need for statistical techniques required for establishing, controlling and verifying process capability and product characteristics." This might well be the only paragraph which does not improve on the 1987 edition. Why must the supplier identify the need for statistical techniques? He might well identify that there is no need for statistical techniques!

the emphasis on documentation. There is little doubt that the committee's desire to emphasize "documented procedures" is well intentioned. Nonetheless, I believe that this desire to emphasize documentation in just about every clause is of dubious value. Indeed, although I do not question the value of documenting certain procedures (corrective action, laboratory procedures, calibration, etc.), I do wonder what is the intent of requiring documented procedures for a contract review, for example. How detailed should the documentation be and for whose benefit? Should there be a generic procedure covering all products, or should there be several generic procedures for several classes of products (e.g., special orders and catalogue orders). Should each product have its own contract review? That could be very tedious indeed.

In such cases, I am not clear as to who is best served by such specific requirements. The customer, the supplier or, could it be the registrars? If the supplier is required to document more and more procedures, are we not in effect giving more and more opportunities for the third-party auditors to audit more and more procedures and, by necessity, find more and more nonconformances. Finally, what are the benefits to society? I leave it to the reader to decide whether or not the need to maintain more documented procedures offers any value added and to whom.

A Look at ISO 9002 and 9003 Updates

Upon reviewing the draft updates to ISO 9002 and ISO 9003 one notices that the paragraph numbering is now the same as in ISO 9001. This is clearly an improvement. The draft standards are also longer: seventeen pages (vs. six pages for the 1987 edition) for ISO 9002 and eleven pages vs. two pages for the 1987 version of ISO 9003. This is mostly due to the fact that most clauses tend to be more detailed. The nature of these clarifications is essentially the same as those explained in Table 9.1. Finally, one should note that the introductory statements are longer (two additional pages for all standards).

One also notices that *Servicing* has been added to ISO 9002 and that *Internal Quality Audits* has been added to ISO 9003. Some paragraphs have been renamed. Document control is now *Document and Data Control* (as in ISO 9001), Product Identification is now *Product Identification and Traceability*, Quality Records is now *Control of Quality Records*.

ISO 9000 and ISO 9004

Both of these documents are substantially longer than the 1987 versions. ISO 9000 is twenty-nine pages long and ISO 9004 is now thirty-nine pages long. Given the scope of this chapter it is impossible to review all of the changes. Basically, both ISO 9000 and 9004 provide the reader with more detailed information. The "Table of Contents" for the ISO 9000 document is provided in Table 9.2. Perusal of Table 9.2 reveals that much more information has been added to the "1992" draft version.

Table 9.2 Table Of Contents for Draft Version of Updated ISO 9000 Document

	Foreword
0.0	Introduction
1.0	Scope
2.0	Normative references
3.0	Definitions
3.1	Terms and definitions taken from ISO 8402 (1987)
3.2	Terms and definitions taken from DIS 8402-1 (1991)
3.3	Additional terms and definitions for the purpose of this International Standard.
4.0	Principal concepts
4.1	Key objectives
4.2	Stakeholders and their expectations
4.3	Facets of quality
4.4	Quality system requirements and product requirements
4.5	Generic product categories
4.6	Concept of a process
4.7	The network of processes of an organization
4.8	Quality system in relation to the network of processes
4.9	Evaluating quality systems
5.0	The role of documentation
5.1	The value of documentation
5.2	Documentation and evaluation of quality systems
5.3	Documentation as support for quality improvement
5.4	Documentation and training
6.0	Quality system situations
7.0	Types of International Standards on quality systems
8.0	Use of International Standards on quality
9.0	Use of International Standards on quality systems for external quality assurance purposes
9.1	General guidance
9.2	Selection of model for quality assurance
9.3	Demonstration and documentation
9.4	Additional considerations in use of quality assurance standards in contractual situations
10.0	Proliferation of standards

Conclusions

It is too early to tell when and how much of the updates will be adopted. Some of the draft documents still have misspelled words. The above pages were meant to be read as advanced "notice" of what is likely to happen within the next couple of years. In their present draft form, the standards do emphasize more documented procedures but, overall, they are a significant improvement from the 1987 version. It remains to be seen what ISO's future will be; this is the topic of our next chapter.

10 The Future of ISO 9000

Above all, standards are a means of communication and of imposing appropriate technical discipline of the operation of the company. [1]

When U.S. corporations first discovered the ISO 9000 series of standards, in 1989, the initial perception, at least within some circles, was that the standard was yet another cleverly disguised barrier to (free) trade. The myth of ISO 9000 led to some early debates and counter debates as to the worthiness of the series.[2] Those involved with the process of developing standards do recognize that there are many economic advantages to adopting national and/or international standards. Robert Toth explains that, although difficult to quantify, the benefits of standardization "are usually directly proportional to the number of activities that are affected by a standard or group of standards."[3] After conducting some forty or more public and in-house seminars to hundreds of participants, I believe it is time to reflect on some important issues surrounding the ISO 9000 registration process and the ISO 9000 series in general. Some of the important issues I would like to address in this chapter are:

- The need to standardize third-party audits.
- The political economy of ISO 9000.
- The dangers of institutionalizing the ISO 9000 series.
- ISO 9000 and litigations.
- ISO 9000 and the innovative process: common sense or contradiction.
- How open is the TC 176 committee?
- Product certification: How complicated?

---

[1] Alan J. Shearer in Robert Toth, editor, *The Economics of Standardization* (Minneapolis: Standards Engineering Society, 1984), p. 46.

[2] See for example Ira Epstein's letter, "More Study Needed on Registration Issue," in *Quality Progress*, 13 (1989). See also James L. Lamprecht, "Demystifying the ISO 9000 Series Standards," *Quality Engineering*, 4 (2), 1991-92, pp. 159-166.

[3] Although Toth's comments were intended for technical standards, the same could be said of quality assurance (systems) standards. Richard Toth, *Economic of Standardization*, op. cit., p. 21.

Third-Party Audits: Can They Be Standardized?

The number of registrars operating in the U.S. has increased significantly ever since 1989 when only a handful of, mostly British, registrars were granting certification. This increase in national and international registrars competing for the lucrative U.S. market, has had positive and negative side effects. On the positive side, customers are beginning to see more competitive daily rates. The reduction in daily rates does not necessarily mean that estimated *total registration cost* has itself reached an equilibrium. Indeed, although daily rates may vary from a (very) low of $800 per auditor/day to a high of $1,500 plus expenses (of course), the number of audit-days, optional pre-assessment visits and other incidental fees such as quality manual review for example, varies greatly; so much so, that it can lead to a difference of up to $15,000 or more, per quotation! Hence, it pays to shop around.

Another important issue to consider is the audit process itself. There is in fact no audit standardization process. This does not of course mean that no attempt has been made to standardize the inherently subjective process of auditing. There is the ISO 10011-1,2 and 3 *Guidelines for Auditing Quality Systems* series and the many approved lead assessor training courses which must be attended by all potential ISO auditors.[1] In addition, most registrars ensure that all auditors receive additional internal training on how to interpret an audit to the various standards. The problem is that each registrar and/or auditor interprets the standards somewhat differently.[2] Consequently, as a customer you are exposed to two sources of variability: between-registrars and within-registrars variability (see Chapter 8). Naturally,

[1] The three standards are Part 1: Auditing, Part 2: Qualification criteria for quality systems auditors and Part 3: Management of audit programs. The standards are available from the ISO New York office (212) 642-4900.

[2] This problem might be mitigated within the next few months with the formation of an association of accredited registrars. See *Quality Systems Update* of November 1992, p. 21 ("Accredited Registrars Consider Establishing Association").

one would hope that the within-registrars variability is reduced as much as possible (mostly via training) but what about between-registrars variability? Can it be reduced? I do not think that will be possible in the near or distant future. One must nonetheless concede that the various attempts by the Institute of Quality Assurance in London and the Registrar Accreditation Board of Milwaukee are designed to control as much as possible the auditing process. However, having had numerous conversations with British, French, Canadian and American registrars, I can assure the reader that although there tends to be a general philosophical agreement on *some* interpretations, not all registrars interpret the standard with the same *rigor*. Naturally, each believe their interpretation to be the correct one. This is of course inevitable.

The likelihood of ever achieving standardization is exacerbated by the fact that guidelines on how to interpret the ISO 9000 series are being produced at an alarming rate. For example, there are at least three sets of guidelines available for the chemical industry. The first, entitled *EN 29001/ISO 9001 Guidelines for Use by the Chemical Industry*, was produced by CEFIC, the European Chemical Industry Council; in my opinion, it is not as good as the *Quality Assurance for the Chemical and Process Industries*, prepared by the American Society for Quality Control. A third set of guidelines, of little substance, was also produced by the U.K.'s Chemical Industries Association.[1] Similar observations can be made within the software industry where several "standards" are available. An Australian standard (AS3563.1-1991) entitled "Software Quality Management System," which incorporates ISO 9001, has apparently been adopted by the IEEE Standards Board. Naturally, such actions are not likely to please members of the ISO work group responsible for the issuance of ISO 9000-3 Part 3: *Guidelines for the Application of ISO 9001 to the Development, Supply*

---

[1] *ISO 9001-EN29001 BS 5750: Part 1:1987*. C.I.A. Guidelines for use by the Chemical and Allied Industries. Chemical Industries Association Limited, London. See also, *Quality Assurance for the Chemical and Process Industries: A Manual of Good Practices*. Published by the American Society for Quality Control, 1987.

and Maintenance of Software. To complicate matters further, certain industry leaders such as Bellcore (Bell Communication Research, Inc.), are rewriting technical reference documents such as the *Software Quality Program Generic Requirements* to conform with both ISO 9001 and ISO 9000-3 (*Guidelines for the application of ISO 9001 to the development, supply and maintenance of software*). Finally, the U.K.'s Department of Trade and Industry has recently introduced the TickIT program which, according to a DTI brochure, "gives the software industry an accredited quality management system certification scheme that meets the special needs of the industry, enjoys the confidence of professional staff, and commands respect from purchasers and suppliers."

Although one could hardly disagree with the objectives set by TickIT, one wonders if other similar industry specific standards (perhaps even more software standards) will be developed by the French, Spanish, Germans, . . . etc. Moreover, if a firm is accredited under the TickIT program, will that certification be recognized in Denmark, or any other European country? What will be the impact on ISO 9000 registration?

For software development firms, the issue as to if or when the ISO 9000-3 Guidelines (or some mutation of TickIT) will become THE international standard for software development has not yet been resolved. From a technical point of view, I find the Bellcore TR-TSY-000179 (*Software Quality Program Generic Requirements*) far superior to the ISO 9000-3 Guidelines. I should however emphasize that, although the two documents do overlap, they were developed to manage two different systems. The Bellcore document focuses on the technical aspects of software quality development, planning, support and control. The ISO 9000-3 Guidelines focuses on *Quality management and quality assurance* and in many respects simply paraphrases the ISO 9001 document. Consequently, the 33 pages of the Bellcore document offer much more technical information than the 15 pages of ISO 9000-3. In the interim, I would suggest that software companies achieve both ISO 9001 registration and Bellcore approval.

This proliferation of industry-specific standards and guidelines is understandable and perhaps even necessary as more and more industries try to adapt the ISO series to their industry. It will nonetheless further confuse the standardization process. Faced with this plethora of standards, users are well advised to first inquire which standard will be recognized before attempting to implement the requirements.[1]

It is imperative that customers seeking registrars, contact as many registrars as possible and ask as many questions as possible. Some questions will still remain unanswered. For example, many customers would like assurance that their ISO registration will be good in all European countries. Such assurances are difficult if not impossible to make simply because European registrars have only negotiated (so far) bilateral agreements and not multilateral agreements (see Appendix G). What then can be said of the political economy surrounding ISO 9000 registration?

The Political Economy of ISO 9000

The ISO 9000 series can find some of its roots in the 1985 Product Liability Directives of the European Economic Community (Official Journal L-210). Although originally developed to ensure a minimum acceptable level of quality assurance, particularly as it relates to so-

---

[1] One of the British Standard Institution's latest "standards" include the BS 7750 on "eco-audits." Of course, environmental audits are not new in this country. The Environmental Protection Agency announced its environmental auditing policy statement on July 9, 1986. Environmental audits have already been performed in California for the last two to three years. The Strategic Advisory Group on Environment (SAGE), a new group within ISO, is currently developing so-called green standards. See J. Ladd Greeno, Gilbert S. Hedstrom and Maryanne DiBerto, *Environmental Auditing: Fundamentals and Techniques*. Center for Environmental Assurance, Arthur D. Little, Inc., 1985. See also *Environmental Audits* (6th edition, October 1989). Lawrence B. Cahill and Raymond W. Kane eds. Government Institutes, Inc., 966 Hungerford Drive, #24, Rockville, MD 20850.

called regulated products (see Chapter 1), there is some evidence that certain European customers seem intent on using the standards as a form of entry barrier (euphemism for protectionism). Such allegations are naturally difficult to prove but they can nonetheless be substantiated with the following accounts (naturally, the companies must remain anonymous).

A U.S. service company with offices in most European countries lost a multimillion dollar European contract because their U.K. offices were not ISO 9002 registered. A U.S. supplier of chemicals used in paint found himself dropped from the list of prefered suppliers because he was not ISO 9000 registered. The Dutch customer opted for a German supplier who was ISO registered. Similar stories can be told about other American suppliers who have been asked whether or not they were considering ISO 9000 registration "in the near future?" Finally, one should mention that in some industries, American firms are finding out that European competitors are marketing/advertising their supposed competitive advantage simply because they are "ISO registered."

One should not however conclude that the above two cases are the rule; in fact they seem, so far, to represent the exception. Many U.S. suppliers still continue to supply to Europe and the rest of the world and have not yet heard from their European or other overseas customers about ISO 9000. The pressure to achieve registration within the next few months is directly related to the type of products your company manufactures. If you are in the business of exporting medical devices or food-related products for example, the pressure to achieve registration as soon as possible is much more urgent than if you are exporting propylene or ethylene. Nevertheless, as ISO gains in popularity, and as more and more industry standards align themselves with ISO 9001, requests for ISO 9000 registration are likely to increase and originate from all over the world. Australians, Argentinians and perhaps even Tunisians may soon request your marketing department to provide them with information regarding your expected ISO 9000 registration date.

In a decade where regional economic alliances are beginning to take shape, one cannot ignore the politico-economic impact of ISO. Although it is still too early to predict which path the North American Free Trade Agreement (Nafta) will take, early signs seem to indicate that the intent of the agreement is to lean towards various forms of protectionism. If that is so, European leaders might then feel justified in retaliating. Among the multitude of tools available for restricting trade, ISO 9000 registration could very well play a small but significant part.[1]

The Dangers of Institutionalizing the ISO 9000 Series

As the process of registering and maintaining registration matures, the whole system runs the risks of becoming blindly and routinely institutionalized much as the MIL Q 9858A and other MIL standards have become. Is that necessarily bad? To some extent yes, because systems tend to ossify unless they are continuously revitalized. A friend of mine who works for a company that still supplies to the defense industry, recently explained how his company, which must comply to many MIL standards, has learned to astutely adapt to the MIL requirements. Over the years, the company in question has developed an outer shell of MIL experts who act as a buffer to ensure that those within the company are protected from the varied requirements imposed by the standards. These experts make sure that on the day of the audit, every item (e.g., hammer, computer, log books, etc.) is where it is supposed to be and people are properly coached on what to say to the auditors. In essence, the company has learned to operate under two systems: one that essentially ignores the standards for the majority of the time; and, one that operates within the constraints of the standards some of the time.

---

[1] See Bob Davis, "Sweetheart Deals: Pending Trade Pact with Mexico, Canada has a Protectionist Air," *Wall Street Journal*, Wednesday, July 22, 1992, pp. 1, 5.

Since the ISO 9000 series very much resembles Mil Q 9858A, one wonders if a similar fate might not await most ISO quality assurance systems. Faced with a plethora of standards, suppliers might be tempted to treat ISO as yet another supplier quality assurance *program* which will be enforced a few weeks prior to the scheduled (and thus routine) audit. In an effort to simultaneously operate under increasingly tight deadlines and with a reduced work force, "lean and mean" suppliers may feel justified in developing an internal set of *informal* guidelines designed to shield workers from the plethora of standards including ISO 9000 (see Figure 10.1). Already, some disturbing signs relating to the "new European rules" have emerged. In a September 7, 1992 *Business Week* article entitled "10,000 New EC Rules", one could read that a supplier had to purchase $5 million worth of high-tech and testing equipment. The article states that "Besides making what it regarded as an unproductive investment, the company had *to add 10 people to its standards staff of six.*"[1] To avoid such costs more and more companies are beginning to invest in software that will facilitate their documentation effort. Starting in early 1991, one could purchase quality manual diskettes for $99 from one East Coast entrepreneur (the price has now dropped to $25). Today, several individuals in the U.S., Canada and even the U.K. advertise similar services designed to facilitate your ISO documentation needs. The software provided by these companies allows for fancy and elaborate document control templates, all you have to do is enter the name of your company, type a few sentences, save the file and presto, you have a fully documented system; but, do you really have a working system? Or, have you in fact simply implemented an informal system designed to "look pretty" but which in fact has very little to do with reality (see Chapter 2). There is certainly nothing wrong with trying to computerize your documentation, and if a software package can make your life easier, why not? However, do try to avoid the temptation of simply "filling in the blanks." To do so *should* attract some reprimand from any registrar.

---

[1] *Business Week*, September 7, 1992, p. 50 (italics added).

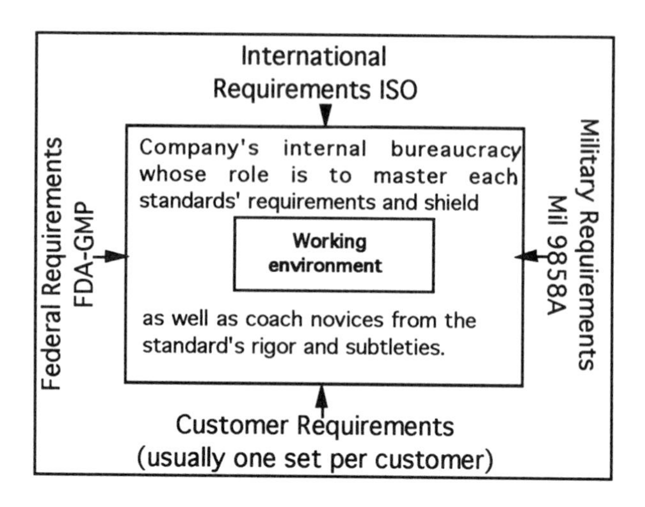

Figure 10.1 Adapting to Constraints

Certainly, ISO 9001 and 9002, have built-in mechanisms designed to prevent the system and its "enforcers" from becoming complacent—namely, internal quality audits. These audits are supposed to ensure that the system remains *effective,* however, it is much too early to know if these audits are indeed performing their tasks. Nevertheless, unless the ISO 9000 series and registrars learn to adapt to the needs of their customers, the series runs a very real risk of being shortcircuited and replaced by yet another "informal" standard.[1]

ISO 9000 and Litigations: Real or Perceived Threat?

Since the EN 29000 series (European equivalent to the ISO 9000 series), is cited in many of the European directives which require the

[1] An analogy could be made with the economic development of some Third World countries. In a book entitled *The Other Path*, author Hernando De Soto explains that the inability of the Peruvian government to develop a modern democratic economic system has forced Peruvians to develop over the years an "informal economy" (hence the title), designed to satisfy basic needs such as transportation, housing and trade; see Hernando De Soto, *The Other Path* (New York: Harper and Row Perennial Library, 1989).

affixing of the CE mark, it is not surprising to learn that there might indeed be some legal aspect surrounding registration.[1] Articles 1 and 2 of the Commission of the European Communities state:

> Article 1 (Scope)
>
> This Regulation lays down rules for affixing the CE mark of conformity provided for in the Community legislation in respect of the design, manufacture, marketing, putting into service and/or use of the industrial products.
>
> Article 2 (Meaning)
>
> 1. The CE mark affixed to the industrial products indicates that the natural or *legal* person who affixed the mark or had it affixed has ascertained that the product conforms to all binding Community provisions applicable to it.
>
> 2. The *legally* binding Community provisions referred to in paragraph 1 are those which cover the products concerned and which provide for appropriate conformity assessment procedures, to the exclusion of any differing national provisions.
>
> 3. Any industrial product covered by the provisions referred to in paragraphs 1 and 2 must bear the CE mark except in cases provided for in the specific directives.[2]

---

[1] Modules D, E and H call for EN 29002, EN 29003 and EN 29001 respectively. See the *Official Journal of the European Communities*, No. I. 380/13 of 31.12.90 entitled, "COUNCIL DECISON of 13 December 1990 concerning the modules for the various phases of the conformity assessment procedures which are intended to be used in the technical harmonization directives."

[2] Commision for the European Communities COM(91) 145 final - SYN 336. "Proposal for a COUNCIL REGULATION (EEC) concerning the affixing and use of the CE mark of conformity on industrial products." Brussels, 17 May 1991, pp. 17-18.

Since the Council regulation cited above will come into effect on January 1, 1993, it is of course too early to tell what opportunities lawyers will find in the ISO/(EN) (2)9000 series. Provisions of the EC product liability law (adopted August 7, 1985) are similar to those of U.S. law and "include holding the manufacturer of a product liable for damage, loss or injury caused by a defect in the product. The directive provides for defenses similar to those available in the United States, including a reference to contributory negligence."[1] There is little doubt that product liability lawyers are closely watching developments within the ISO 9000 arena; in fact, some lawyers are already hard at work studying the many volumes of documents produced by the European Commission. It might not be too long before lawyers ask companies to produce their tier one or tier three documents. The ensuing nightmare of legal and para-legal verifications may well reach gargantuan proportions as registrars are next asked to justify how a particular company could have been granted a particular certificate when in fact fastener A3-25 or a 3/4 inch bolt failed in the field?[2]

ISO 9000 and Innovation: Common Sense or Contradiction?

The need to continuously improve and innovate in order to remain competitive has long been maintained by management specialists such as Peter Drucker, economists such as Michael Porter and quality gurus such as Tom Peters. Since innovation, creativity and continous improvement are important variables in the search to achieve and maintain competitiveness, it is reasonalbe to ask what role, if any, the ISO 9000 series might have in helping or stifling a firm's competitive stance.

---

1 "EC Single Market Law Affecting Exporting and Distribution," published by the United States Department of Commerce, International Trade Administration, December 1, 1991.

2 Most registrars seem to have already anticipated such difficulties by including in their contracts clauses absolving them and their auditors of any legal action.

Tom Peters, one of the latest proponents of the innovation school, states in his *Thriving on Chaos* that, "[T]he Flexible, Porous, Adaptive, Fleet-of-Foot Organization of the Future: Every Person is "Paid" to be Obstreperous, a Disrespecter (sic) of Format and Boundaries, to Hustle and to Be Fully Engaged with Engendering Swift Action and Constantly Improving Everything."[1] In his characteristically upbeat and enthusiastic approach, Peters concludes his book with the following comments:

> Thus, we can readily envision an astonishingly high degree of "controlled" flexibility and informality, starting with the front line and outsiders, in our "new-look" organization. But there is also an astonishing amount of hard work required—e.g., perpetually clarifying the vision, living the vision, wandering, chatting, listening, *and* providing extraordinary and continuous learning opportunities—that must precede and/or accompany all this. So perhaps "purposeful chaos" is the best description of the new-look firm.[2]

I wholeheartedly agree with the concept and, in fact, most of the companies I have visited do appear to operate in a self-imposed stressful (rather than purposefully) chaotic mode.

I once visited a company which took great pride in its "concurrent engineering" process. Although not an expert on the topic, I came to find out that the process was enthusiastically embraced by the engineering department for it allowed for a considerable reduction in development cycle time (so, I was told). For new product

---

[1] Tom Peters, *Thriving on Chaos* (New York: Harper Perennial, 1987), pp. 659, and 661, Figure 23. It is often difficult to distinguish between Peters' comments and Alvin Toffler's penetrating comments. See for example Toffler's *Power Shift* (Bantam Books, New York, 1990).
[2] Peters, op. cit., p. 666.

development, the company in question would develop a preliminary design and start production, not on a prototype, but on an "experimental production unit" which would naturally go through a variety of upgrades and improvements before being finally released for **shipping!** The process is actually not very different from what I had often witnessed and have come to refer to as "just in time engineering".[1] What is interesting in this case and, no doubt, many others like it, is that the company has been successful for several decades and is a leader within the industry. This is apparently one of the only ways the company believes that it can respond quickly enough to market/customer demands. By so doing, it has created for itself a successful niche. Would such a company pass an ISO 9001 audit? Not likely. Would the implementation of an ISO 9000 type quality assurance program endanger its ability to rapidly adapt to market needs? This was one of the primary concerns of one of the V.P.s. What about "controlled flexibility and informality" and "purposeful chaos," aren't these concepts the very antithesis of the ISO 9000 series? Is a quality assurance system designed along ISO requirements conducive to the implementation of "swift actions?" Can a system which emphasizes the approval and proper documentation of procedures tolerate anyone who is "paid to be obstreperous" and a "disrespecter (sic) of formal boudaries?" Not likely; and yet, activities encouraged by Peters could be maintained within an ISO 9000 organization as long as they are not within the system or at least transparent to the system.[2] The topic, intriguing as it is, is too complex to be covered presently.

---

[1] I do realize that some purists might be appalled at the notion of just in time engineering. Here is not the place to debate its virtues, or lack thereof. What is actually very distressing about this particular case is that the 'final' product is released under the same 'preliminary' drawings! "Where are the revisions?" you might ask. Where indeed?

[2] I will let the reader ponder those questions. I should explain that although the system was praised by the engineering department, the manufacturing director did not much care for the informality. As the reader can well imagine, the system created great stress for manufacturing as well as other (notably materials) departments.

How Open Is the TC 176 Committee?

The *ISO 9000 News* published in May 1992 a revealing interview ironically entitled "The invisible architects of quality systems." During the course of the interview, Reg Shaughnessy and Peter Ford, respectively Chairman and Secretary of the ISO Technical Committee 176 (ISO/TC 176) revealed some interesting point of view. Highlights of the seven page interview are summarized below (italics have been added to the answers for emphasis).[1]

ISO 9000 News: *Can anyone volunteer to join TC 176 subcommittees and their working groups?*

Reg Shaughnessey: It really is up to the leader of the national delegation present at the subcommittee level to decide in principle who can join the work sessions, and why.

We have however a problem at the working group level which is, in principle, composed of *assigned experts.* As a subject becomes popular, many national bodies, national business or industry leaders volunteer to participate. In certain working groups, too many people in the group sometimes are there just to learn, not really to contribute. The negative aspect is that the working group then becomes a forum for debate instead of a meeting of *professionals* producing the drafts of working documents.

ISO 9000 News: *What can you do?*

Reg Shaughnessy: In future, we intend to ensure that the only people participating in working groups are those who have been *appointed* by the subcommittees. Others will be granted "observer" status and

---

[1] *ISO 9000 News: A Newletter on Quality Management Standards*, Vol. 1, No. 3, May 1992, pp: 4-5.

assigned to a sort of business gallery. *Their voice will only be heard through their delegates meeting in plenary session.*

To continue our work effectively and efficiently, we must control the access to debate at the working group level.

ISO 9000 News: *How can this be achieved?*

Peter Ford: . . . National member bodies in conjunction with the subcommittee are responsible for putting experts on specific groups . . . . We want to have the experts at the table and put all learners around the edge so that the working group can get on with its work.

ISO 9000 News: *Do you get direct feedback from industry?*

Reg Shaughnessy: *No,* in the sense that member bodies are in charge of collecting and synthesizing the feedback from their industry . . .

Peter Ford: The way it has all been set up is that input comes through the delegation. It is up to the individual nation to make sure that their delegation, and its input, is correctly put together, and comprehensive.

ISO 9000 News: *The need of a worldwide recognition of accreditation seems to be growing namely for reasons of efficiency. What respective role do you see for the ISO 9000 Forum and for the 10000 series of standards?*

Reg Shaughnessy: The question is an interesting one. TC 176 is merely writing the prescription for the system. This having been done, somebody will come forward and say, "But this is a global perspective. Who drives it?" Our reply will be: "Not us" because our mission is foremost to write the prescription, neither to implement it nor to check out its implementation. We are writing the rules, not playing the game . . . and certainly not refereeing it.

It is the author's belief that the spirit of isolationism reflected in Mr. Shaughnessy's comments, although understandable to some extent, are nonetheless not likely to help the ISO 9000 process in the long run. The fact that "experts" have assumed the responsibility to write a standard without apparently being concerned as to its feasibility or ease of implementation is, to say the least, disturbing. I leave it to the reader to arrive at his/her own verdict.[1]

Finally, I would like to conclude this overview by presenting some issues relating to product certification.

Did You Say Product Certification?

If you thought that achieving ISO 9001, 2 or 3 registration was going to solve all or most of your exporting problems, think again. The recent (1992) publication of *Certificat: Product Certification European Directory*, published by the AFNOR (the French Standard Association), complicates matters somewhat. The nearly 700 page trilingual document (English-French-German) is subdivided into five chapters: 1. Certification in Europe, 2. Description of certified products by country, 3. Certification bodies by country, 4. Certified products by certification system and 5. Certified products by sector of activity.

Reading the certification program for each country (E.C and E.F.T.A), one notices that most countries mention either the EN 29000 (ISO 9000) series of quality assurance systems or the ISO 45000 series for notified bodies.[2] As one scans the 700 page document, it becomes evident that headings are not standardized across countries. For

---

[1] The situation is even worse in the ISO/TC 69 committee which addresses statistical issues. The majority of these standards address esoteric topics and are written by academics *for* academics. Similar comments could be made about the "new" environmental standards which emphasize the managerial aspects of so-called "environmental" quality systems and are written with little if any input from environmentalists, biologists, hydrologists, geologists, biochemists, etc.

[2] The countries are: Denmark, Finland, France, Ireland, Greece, Italy, Norway, Portugal and the United Kingdom. Surprisingly, Belgium does not cite the EN 29000 series!

example, whereas Italy has a heading entitled "Internationally recognized standards", the Netherlands list "EEC type examination," France has "homologation" and Greece has two categories referred to as "System ISO no. 1 and System ISO no. 5."

One also notices the great disparity in the number of certification bodies between countries. Thus, whereas some countries such as Spain lists only one certification body, namely the AENOR, Germany lists approximately 400 certification bodies and trade organizations; the most for any country!

It gets more complicated. Suppose you are manufacturing respiratory equipment and you would like to know what product certification(s) will be required for exporting to Europe. You might think that if you achieve ISO 9001 certification your might be able to insert the "CE" mark on your product and export to Europe. Well, it might unfortunately be a bit more complicated. If you export to France you will find a product heading called "Cardio-respiratory monitors for home use." Suppose, for the sake of this argument that this category happens to fit your product.[1] The basic mandatory specification for France is listed as L.665-1 and R.5274. What would Germany require? There is unfortunately no "medical equipment" heading for Germany. Instead, the nearest equivalent for Germany is called "Safety and Health." Under that heading, you will find no equipment listed under "cardio-respiratory"! If you look under "respiratory," you will find eight headings, the closest being "Respiratory apparatus." Assuming that this category approximately corresponds to the French heading, you will need to obtain a GS mark which could be obtained from one of three registration bodies as opposed to only one in France.[2] If you would like to export to Spain, you will pleasantly be surprised to find out that Spain does not even have a heading for medical or health

---

1 In most cases you will find that it is in fact difficult to find the corresponding product heading.

2 In Germany, the GS marking system is based on legislation concerning industrial technology. Germay has six "framework" systems known as A-F. See *Certificat*, p. 8.

apparatus! Of course, should you want to export your product to all Western European countries, you might have to invest at least three to four hours just to thumb through the *Certificat*.[1] The reader can perhaps now sympathize with the various technical harmonization committees meeting in Brussels. The task is enormous and, to my knowledge, no one individual has yet *the* answer, which is not likely to unfold for the next couple of years.

Will the ISO 9000 Series Make It to the 21st Century?

I don't really know. Some enthusiasts seem to naively believe that the 9000 series is here to stay, forever! But why should it survive the test of time? Perusing several chapters of Alvin Toffler's *Power Shift*, I found myself re-evaluating some earlier position I had with respect to standards in general. In a chapter entitled "The Widening War" (Chapter 12), Toffler devotes a few pages to the effect of technical standards.

> As business produces more diversified products, there is, in addition to a mounting pressure for more standards, a countereffort to make products more and more versatile by accommodating multiple standards.
>
> Standards have long been set by industries or governments to assure the safety or quality of products and, more recently, to safeguard the environment. But they are also designed by protectionist governments to keep competitive foreign products out or to advance an industrial policy.[2]

---

[1] It is interesting to note that no reference is made under the rubric "Europe" where European directives are listed!

[2] Alvin Toffler. *Power Shift*. (Bantam Books, New York, 1990), pp. 138-139.

Certainly, the ISO 9000 series is not a technical standard nor is it a national standard, however, the phenomena of proliferation and protectionism are currently unfolding. The use of ISO 9000 certification as a means to obtain (or deny) contracts, has already been stated above. As for proliferation, examples of ISO 9000 penetration within the software industry, environmental audits and educational institutions, have already been mentioned throughout the book.

Assuming that the proliferation of ISO 9000 type standards is of no consequence; can we then assume that the ISO 9000 series is ideally suited for companies that will have to survive the *power shift* required to survive into the next century? Toffler (and apparently Peters) believes that companies will have to introduce "corporate glasnost - an openess to imagination, a tolerance for deviance, for individuality, . . .".[1] In additon, information will no longer be controlled by specialists (cubbyholism). Rather, channels will have to be opened to allow for the free, and rapid, flow of information:

> Today, high-speed change requires high-speed decisions—but power struggles make bureaucracies notoriously slow . . . but bureaucracies try to eliminate intuition and replace it with mechanical, idiot-proof rules. . . This explains why millions of intelligent, hardworking employees find they cannot carry out their tasks . . . except by going around the rules, breaking with formal procedures.[2]

Will an ISO 9000 type quality assurance system, which emphasizes documented procedures (often controlled by various departmental bureaucracies), help companies become "flex-firms"? Moreover, how can companies implement a quality assurance system which will allow them to do simultaneously what they used to do sequentially? The challenge needs to be addressed. I do not believe that the ISO 9000

---

[1] Toffler, op. cit., p. 152.
[2] Toffler, op. cit., p. 178-179.

series, in its current state, addresses the issues raised by Toffler and others.[1] In fact, I believe that the ISO 9000 series, as it is currently being implemented by many companies, contradicts *most* of what Toffler, Peters and others have to suggest. However, all is not lost, for indeed, one of the major benefits of the ISO 9000 series is that it should at least encourage departments to work together as they help their company achieve ISO 9000 registration. Moreover, if done properly, communication should also improve between departments. Finally, if it is to survive well into the next century, the ISO 9000 series will need to adapt. Failure to do so will no doubt considerably reduce its chances of survival.

Conclusions

The flurry of ISO 9000 activity which started in the U.S. in 1990 has generated some criticisms. At issue is whether or not the ISO 9000 mania will survive the 90s. British author Allan Sayle does not seem to think so. The major concern expressed by Sayle and others is that too many audits have become a paperwork hunt designed to ensure, in part, that calibration stickers are placed on gauges.[2] This "dot and sticker myopia," which I have often witnessed, is certainly of little value to management and the company as a whole.

Companies wishing to implement a Total Quality Management (TQM) philosophy will perhaps be disappointed to find out that the ISO 9000 series does not, in its present form, deliver a comprehensive TQM system. Such criticisms are, however, unfair to the ISO 9000 series, which, after all, only addresses quality assurance systems. Moreover,

---

[1] See for example Tom Peters latest book, *Liberation Management*. (Alfred A. Knopf: New York, 1992) and William H. Davidow and Michael S. Malone, *The Virtual Corporation* (Edward Burlingame Books/Harper Business: New York, 1992)

[2] Mr Sayle's speech was presented at the 1st Annual Quality Audit Conference in St. Louis, Missouri. A summary of the speech (including a rebuttal by George Lofgren of the RAB) can be found in the Quality Systems Update newsletter (Volume II, No 3, March 1992). The reference to ISO 9000 mania is from a speech made by Frank R. Gollhofer, "Will API Licensing Program Withstand ISO 9000 Mania?", Symposium on Quality System held during the API 68th Annual Standardization Conference 1991.

one should remember that the ISO 9000 series finds its roots in military standards developed in the late 1950s. These origins do, to some extent, explain the anachronisms found in some of the clauses of the ISO 9000 series.

Although the ISO 9000 series is rapidly gaining popularity, it is still far from being universally accepted in the United States. During one phone conversation with a potential customer who supplies to the aircraft industry, the subject of ISO 9000 came up. Although the individual had heard of ISO, he felt there was no need to implement such a system at his company. The individual went on to explain, with some relish, how one local supplier had recently told Airbus to "go pound sand" when asked about ISO. Such arrogance might bring some short term pleasure, however, the fact remains that it is not likely to help business in the long term. A report published by the Aerospace Industries Association warns that:

> Today, in standardization as in other fields, the U.S. is no longer the unquestioned world leader, but a strong player among strong rivals. Standards developed outside of the United States—particularly in Europe or in international standards organizations—are gaining credibility and acceptance. Key examples are the Joint Aviation Regulations (JARs) developed in Europe, and the ISO 9000 series on quality systems developed by the International Organization for Standardization. To the extent that these standards diverge from or conflict with U.S. standards and practices, the U.S. can be at a disadvantage.[1]

---

[1] "Impact of International Standardization and Certification on the U.S. Aerospace Industry," Aerospace Industries Association, 1250 Eye Street, N.W. Washington, D.C. 20005. See also Michael Prowse, "Is America in Decline?" in *Harvard Business Review*, July-August, 1992, pp. 34-45 as well as Henry R. Nau's *The Myth Of America's Decline* (New York: Oxford University Press, 1990).

I naturally made sure that the supplier in question received a copy of the AIA report.

Despite its flaws, it is hard to imagine that the ISO 9000 series will easily succumb to severe criticisms and simply wither away within the next three to five years (nonetheless, that scenario could develop). One must recognize that "[I]nternational standards affect national standards, international trade and even national laws and regulations."[1] That is certainly true for the ISO 9000 series. Already its impact has been felt throughout American industries and is likely to be felt until 1994.[2] The vigilance of TC/176 committee—the international committee in charge of updating the ISO 9000 series—will probably do whatever it can to ensure that the ISO 9000 series adapts and thus survives well into the next century. Already, many improvements have been recommended by the U.S. Tag committee on ISO 9000. Whether or not these improvements—to be included in the 1996/1997 updates—will be approved by the TC/176 committee remains to be seen. Meanwhile, companies wishing to increase or even maintain their European (or global) market niche must seriously consider achieving ISO 9000 registration as soon as possible. Failure to do so may indeed lead to some unpleasant and unnecessary surprises.

---

[1] Toth, *Economics of Standardization*, op. cit., p. 21. See also R. Scott's suggestion for an ISO 9000 quality tax incentive in the Letters section of *Quality Progress*, July 1991, p. 13. The Standards Council of Canada does integrate ISO/IEC standards within the various Canadian standards.
[2] See *Business Week* of October 19, 1991, "Want EC Business? You Have Two Choices," pp. 58-59.

Afterword

A few weeks after having written the "last" sentences, I had time to reflect on the contents of the preceding ten chapters. I would not have thought about writing an afterword were it not for the fact that I had accidently come across a most interesting book (interesting, that is, as it relates to the ISO 9000 series). Indeed, while looking for a book by Russell L. Ackoff (*Management in Small Doses*), my eyes locked on an intriguing title, *Systems and Procedures*, by Victor Lazzaro.[1] The first thing I noticed about the book was that it was published in 1959. As I began scanning its Table of Contents, I could not believe what I was reading: Chapter 1 *Systems and Procedure*, Chapter 4 *Systems Charting*, Chapter 5 *The Management Audit*, Chapter 8 *Forms Designs and Control*, Chapter 9 *Records Management*, Chapter 10 *Company Manuals*, etc. Perusing each chapter, I was pleasantly surprised to discover that, although some of the information was dated, much of what the authors had to say, was still very relevant today and in fact, in a strange way, paralleled most of what I and others had "re-discovered" over the years.

It would be impossible to summarize the many interesting passages in a few paragraphs; however, since I suspect readers might have a difficult time finding the book, I have extracted a few of the most pertinent and interesting comments. As you read some of the quotations, remember that the book was published over 33 years ago by individuals who had been practicing their skills for approximately twenty to thirty years. It is therefore important to realize that much of what is proposed within the chapters of *Systems and Procedures*, was thought of and practiced during the 30s and 40s! All quotations are from *Systems and Procedures*.

> Systems and procedures is "the analysis of corporate policies, procedures, forms, and equipment in order to

---

[1] Victor Lazzaro, editor. *Systems and Procedures : A Handbook for Business and Industry* (Englewood Cliffs, N.J. Prentice-Hall, Inc.: 1959).

simplify and standardize office operations." [John W. Haslett, p.12.]

We often hear the very generalized and theoretical statement that "systems and procedures are in some measure the responsibility of every person in any company or agency." Whether or not any individual accepts or rejects this theory seems to be entirely dependent upon how the general statement is translated into specific terms and the extent to which 'every person' gets a clear and plausible explanation of what the phrase 'in some measure' means to him wherever he may be . . . [in the organization] . . . *It is equally apparent that when this clarification of responsibility does not take place, one or both of two things happen: (1) People do little or nothing about systems and procedures, or (2) people who do something establish overlapping, conflicting, or shortsighted objectives.* [William A. Gill, pp. 6-7 emphasis added.]

The successful business executive of today has recognized that management auditing is a tool of management, instrumental in examining and determining the quality of performance. It is also an instrument for measuring the *effectiveness* of a company's organization structure, its policies and practices, its systems and procedures, and its personnel. The proper use of this instrument can be the means of equipping business for better achievement of predetermined objectives. It can also do much, through review and appraisal, in finding improved methods of reducing costs and increasing profits. [William P. Leonard, p. 123, emphasis added.] Still good advice today!

The job of manual development is looked upon as one of keeping key personnel informed of changing management attitudes rather than that of doing a one-time job of drafting the organizational chart and putting policies and

> procedures into permanent book form . . . (p. 248) It is recommended that a copy of the entire organizational manual be provided for every member of management from first level of supervision to members of the board of directors . . . (p. 256) Many companies employ one or a combination of the following manual revision control techniques:
>
> 1. Direct that written procedures be followed without exception, but make it the responsibility of the manual holders to point out the need for revisions when operating experience indicates that current instructions are impracticable.
> 2. Establish regular audit schedules to determine if current practices coincide with written procedures . . . (p. 264)."
>
> [Continues with three more suggestions. James G. Hendrick, pp. 248, 256 and 264.]

After reading the above quotations one should have a very good appreciation of the French proverb *The more things change the more they remain the same* (Plus ça change plus c'est la même chose!).

No doubt, the contents of countless more books similar to *Systems and Procedures* currently await to be re-discovered. It may well be that 25 to 30 years from now, quality professionals—or their nearest equivalent, who might be some mutation of the process systems engineer of the 50s—will re-discover the works published in the 80s and 90s. It would be interesting to find out how the ISO 9000 series evolves, or if it even survives past the next century. As things currently stand, many companies which have recently achieved ISO 9000 registration are quick to point out how the ISO 9000 series saved them from damnation—one wonders how they managed to survive all these B.ISO (Before ISO) years. Will they still have an *effective* quality assurance system five, three or even two years from now? Will they have found more ways to optimize their operations by further reducing their work force? If so, who will manage those quality assurance systems? Word processors? It is too early to tell, but time will tell.

Bibliography

Russell L. Ackhoff. *Management in Small Doses.* New York: John Wiley & Sons, 1986.

Aerospace Industries Association, "Impact of International Standardization and Certification on the U.S. Aerospace Industry," AIA, 1250 Eye Street, N.W. Washington, D.C. 20005.

Karl Albrecht and Ron Zemke. *Service America!* New York: Warner Books, 1985.

ANSI/ASQC Q90 ISO 9000 Guidelines. Milwaukee, Wisconsin, ASQC Press, 1992.

Automotive Industry Action Groups (AIAG), *Measurement Systems Analysis. Reference Manual.* October 1990. 26200 Lahser Road, Suite 200, Southfield, MI 48034.

C. Argyris and D. H. Schon. *Organizational Learning; A Theory of Action and Persepctive.* Reading, MA: Addison-Wesley, 1978.

Rich Arons. *Euromarketing.* Chicago: Probus Publishing Company, 1991.

Donald L. Bartlett and James B. Steele. *America: What Went Wrong?* Kansas City: Andrews and McMeel, 1992.

Walter H. Boehling, "Europe 1992: Its Effect on International Standards," *Quality Progress,* July 1990, pp. 29-30.

Harvey J. Brightman. *Problem Solving: A Logical and Creative Approach* (College of Business Administration, Georgia State University. Atlanta, Georgia: 1980).

Martin M. Broadwell. *The Supervisor and On-The-Job Training.* New York: Addison-Wesley Publishing Company, Inc., 1991.

Bureau of Business Practice. *Bringing Out the Best In Your People.* BBP Professional Information Group of Simon & Schuster, 1992.

Lawrence B. Cahill and Raymond W. Kane (eds.) *Environmental Audits* (6th edition, October 1989). Government Institutes, Inc., 966 Hungerford Drive, #24. Rockville, MD 20850.

Certificat: Product Certification European Directory. Paris: AFNOR, 1992.

Commission for the European Community COM (91) 145 final, Brussels, 17 May, 1991.

William H. Davidow and Michael S. Malone. *The Virtual Corporation.* New York: Edward Burlingame/Harper Business, 1992.

Bob Davis, "Sweethart Deals: Pending Trade Pact with Mexico, Canada has a Protectionist Air," *Wall Street Journal,* July 22, 1992, pp. 1, 5.

Hernando De Soto. *The Other Path.* New York: Perennial Library, Harper & Row Publishers, 1990.

Timothy M. Devinney and William C. Hightower. *European Markets after 1992.* Lexington: Massachussetts, Lexington Books, 1991.

John R. Dixon. *Design Engineering: Inventiveness, Analysis, and Decision Making.* New York: McGraw-Hill Book Company, 1966.

Alan S. Fisher. *CASE: Using Software Development Tools.* New York: John Wiley & Sons, Inc., 1991.

Burleigh B. Gardner. *Human Relations in Industry.* Chicago: Urwin, 1945.

J. Ladd Greeno, Gilbert S. Hedstrom and Maryanne DiBerto. *Environmental Auditing: Fundamentals and Techniques.* Center for Environmental Assurance, Arthur D. Little, 1985.

James M. Jenks and John M. Kelly, *Don't Do, Delegate!* New York: Ballantine Books, 1985.

J.M. Juran and Frank M. Gryna, Jr. *Quality Planning and Analysis.* New York: Mc Graw-Hill Book Company, 1980.

P. Hersey and K. H. Blanchard. *Management of Organizational Behavior.* Englewood Cliffs: Prentice-Hall, 1977.

David A. Hounshell. *From the American System to Mass Production, 1800-1932.* Baltimore: The John Hopkins University Press, 1985.

Daniel Kasprzyk, et al., editors. *Panel Surveys.* New York: John Wiley & Sons, 1989.

David Kersey. *Portraits of Temperament.* Del Mar, California: Prometheus Nemesis Book Company, 1991.

David Kersey and Marilyn Bates. *Please Understand Me: Character & Temperament Types.* Del Mar, California: Prometheus Nemesis Book Company, 1984.

Thomas Kuhn. *The Structure of Scientific Revolutions.* Chicago: University of Chicago Press, 1970, 2nd ed..

James Lamprecht. *ISO 9000: Preparing for Registration.* New York: Marcel Dekker, 1992.

James Lamprecht, "ISO 9000 Implementation Strategies," in *Quality*, November 1991, pp. 14-17.

James Lamprecht, "Demystifying the ISO 9000 Series Standards," In *Quality Engineering*, Vol. 4 (2), 1991-1992, pp. 159-166.

James Lamprecht, "The ISO Certification Process," in *Quality Digest*, Vol. 11, No. 8, August 1991, pp. 61-70.

James Lamprecht, "Consulting and the ISO 9000 Series: Who Needs It?, *1992 Rocky Mountain Quality Proceedings*, June 8,9, 1992, pp. 334-344.

Victor Lazzaro, editor. *Systems and Procedures : A Handbook for Business and Industry.* Englewood Cliffs, N.J., Prentice-Hall, Inc.: 1959.

Don Linville, "Exporting to the European Community," *Business America*, February 24, 1992, pp. 18-20.

John W. Locke, "Quality Standards for Testing Laboratories," in *American Association for Laboratory Accreditation*, 656 Quince Orchard Road #304. Gaithersburg, MD 20878-1409.

Ernest J. McCormick and Daniel R. Ilgen. *Industrial Psychology.* Englewood Cliffs, New Jersey; Prentice-Hall, 1986.

Rosabeth Kanter Moss. *The Change Masters: Innovation and Entrepreneurship in the American Corporation.* New York: Simon and Schuster Touchstone Book, 1988.

Henry R. Nau. *The Myth of America's Decline.* New York: Oxford University Press, 1990.

David F. Noble. *America By Design.* New York: Oxford University Press, 1977.

David F. Noble. *Forces of Production: A Social History of Industrial Automation.* New York: Oxford University Press, 1986.

Donald A. Norman. *The Design of Everyday Things.* New York: Doubleday Currency, 1988.

George S. Odiorne. *The Change Resisters.* Englewood Cliffs, New Jersey: Prentice-Hall, Inc., 1981.

Official Journal of the European Communities, various issues.

Jagdish Parikh. *Managing Your Self.* Oxford: Basil Blackwell, 1991.

V. L. Parseghian. *This Cybernetic World of Men, Machines, and Earth Systems.* New York: Anchor Books, 1973.

C. Robert Pennella. *Managing the Metrology System.* Milwaukee, Wisconsin: ASQC Quality Press, 1992.

Charles B. Perrow. *Organizational Analysis: A Sociological View.* Belmont, CA: Brooks/Cole Publishing Company, 1970.

Tom Peters. *Thriving on Chaos.* New York: Harper Perennial, 1991.

Tom Peters. *Liberation Management.* New York: Alfred A. Knopf, 1992.

Henry Petroski. *To Engineer is Human.* New York: Vintage Books, 1992.

Juliet B. Schor. *The Overworked American: The Unexpected Decline of Leisure.* New York: Basic Books, 1992.

C.J. Skinner, D. Holt and T. M. F. Smith. *Analysis of Complex Surveys.* New York: John Wiley & Sons, 1989.

P. H. Sydenham. *Measuring Instruments: Tools of Knowledge and Control.* London: Peter Pergrinus Ltd., 19879.

Alvin Toffler. *Power Shift.* New York: Bantam Books, 1990.

Robert Toth, editor. *The Economics of Standardization.* Minneapolis: Standards Engineering Society, 1984.

Edward Yourdon. *Decline & Fall of the American Programmer.* Englewood Cliffs: Yourdon Press, 1992.

Paul Watzlawick and others. *Change: Principles of Problem Formation and Problem Resolution.* New York: W.W. Norton & Company, Inc., 1974.

Keith D. Wilcock. *The Corporate Tribe.* New York: Warner Books, 1984.

Appendix A: A Look at Some Directives—Telecommunications, Medical Devices and Food and Beverages

U.S. suppliers to the European market must realize that many of the directives are not likely to ever affect them. Directives such as Directive 62: "Coordination of laws relating to legal expenses insurance"; Directive 73 "Proposal for regulation for a European Economic Interest Grouping"; or Directive 211 "Directive on eradication of Brucellosis in sheep and goats" are nor likely to impact many, if any, U.S. exporters. Such is not the case however for telecommunications, medical and food and beverage directives.

Telecommunications[1]

The European Telecommunications Standards Institute (ETSI) is currently developing 300 *voluntary* standards known as European Telecommunications Standards (ETSs). In some cases, ETSs can be brought together to create a new document entitled Normes Européenes de Télécommunications (NET). *When a NET is approved it becomes a mandatory standard of the European Community.*

The telecommunications terminal equipment directive requires manufacturers to rely on ISO 9000 quality control procedures. Other telecommunications devices affected include: satellite services, mobile communications and telecommunications services. At present, sixteen telecommunications-specific directives and nine proposals exist. A sample of telecommunications directives includes: "Coordinated introduction of public pan-European cellular digital land-based mobile communications (L-69)," "Radio interferences Electro Magnetic Compatibility (L-127)," "Frequency bands reserved for pan-

---

1 References for the next three paragraphs are taken from the following documents: "EC Telecommunications," October 1, 1991, "Medical Devices," April 1, 1992 and "The European Community's Policy and Regulations on Food and Beverages," November 1, 1991, published by the U.S. Department of Commerce. International Trade Administration.

European public radio paging (L-261)," "Protection of computer programs (L-279)," and more.[1]

Medical Devices

There are three directives affecting the medical device sector: active implantable medical devices (AIMD); medical devices directive (MDD), which covers the majority of all active and non-active devices; and in vitro diagnostics directives (IVD), which covers reagents and test kits. European and American Good Manufacturing Practices (GMPs) have incorporated the ISO 9001 model within their GMPs.[2] In order to place products on the European market, U.S. medical device manufacturers will need to go through the necessary steps to *certify that their product conforms to EC requirements by affixing a CE mark to their products.* Manufacturers of AIMDs and MDDs will likely have to implement an ISO 9001 quality assurance system.

---

1 See "EC Telecommunications," Oct. 1, 1991 from the United States Department of Commerce International Trade Administration.

2 In the *Suggested Changes to the Medical Device Good Manufacturing Practices Regulation*, U.S. Department of Health and Human Services (November 1991), one can read the following: "While not the primary reason for wanting to change the GMP, implementation of the suggested changes would result in a GMP which, while not indentical. would be equivalent to the proposed European Community (EC) GMP which is based on ISO 9001." p. 4.

There are currently (in 1992) eight directives concerning medical devices. They are:

#	Directives	Official Journal
L119	Community action in the field of information technology and telecommunications applied to health care—Advanced Information in Medicine.	L134
L212	Directive on the approximation of the laws of the member states relating to active implantable medical devices.	L189
C328	Proposal for a Council Directive concerning medical devices.	C237
----	In Vitro Diagnostics Devices; Working Document: Elements for the forthcoming legislative harmonization of in-vitro medical devices.	
L127	Radio interferences Electro Magnetic Compatibility.	L139
L245	Council Decision concerning the modules for various phases of conformity assessment procedures.	----
-----	Proposal for a Council Regulation concerning the affixing and use of the CE mark of conformity on industrial products COM (91) 145-SYN 336.	
	Commission Green Paper on the Development of European Standardization: Action for faster technological integration in Europe (90/456/EEC).	

Food and Beverages Directives

There are at present, in the 1992 Single Internal Market Program, more than 50 regulations pertaining to food and beverage issues. With respect to food packaging requirements, the directives *should* allow U.S. food manufacturers and packaging agents to follow fewer EC regulations (hopefully one rather than twelve rules). As environmental concerns are becoming increasingly important, directives are curently being implemented to address those concerns. Germany, for example, has introduced a controversial packaging law which would require all packaging used in the transportation of goods to be collected by the retailer. The law is currently being challenged.

New labelling directives have been introduced in recent EC directives. Labels must contain:

1. The name under which the product(s) is sold
2. The list of ingredients in languages expected to be easily understood by national customers
3. The net quantity
4. The "use by" date or minimum durability
5. Any special storage conditions or conditions of use
6. The name or business name and address of the manufacturer or the packager, or of the seller established within the Community
7. Place of origin, especially where labels might mislead the consumer
8. Instructions for use when it would be impossible to make appropriate use of the foodstuff in the absence of such instructions

The new lot marking directive (89/396/EEC) which requires the placement of the letter "L" on the package is likely to impact the U.S. food sector. The twenty-seven directives directly affecting the food industry are too numerous to list here. The directives cover: preservatives, emulsifiers, coffee and chicory extracts, alcoholic strength, frozen foods, food additives, jams, fruit juices, etc.[1]

To obtain information on how to do business in Europe and ask questions regarding EC 1992 call the Commerce Department's Single Internal Market Information Service (SIMIS) at (202) 377-5276. The National Institute of Standards and Technology (NIST) also publishes many articles and pamphlets on the ISO 9000 series. To obtain copies write to: The Standards Code and Information Program, A-629, Administration Building, National Institute of Standards and Technology, Gaithersburg, MD, 20899. The NIST TRF Bldg. (Room A163) can be reached at (301) 975-4040. The U.S. Department of Commerce International Trade Administration publishes *Business America* (a monthly) and *Europe Now* a quarterly newsletter. Write to: Government Printing Office, Superintendent of Documents, Mail Stop SSOM, Washington, D.C. 20401.

---

[1] For further information see, "The European Community's Policy and Regulations on Food and Beverages," November 1, 1991. United States Department of Commerce International Trade Administration.

Additional information is available from:

National Center for Standards and Certification Information (NCSCI) (see NIST for address)

and from

Office of EC Affairs
International Trade Administration, Room 3036
14th and Constitution Ave., SW
Washington, DC 20230
Phone: (202) 377-5276

To obtain the latest draft standards of the European Committee on Standardization (CEN), the European Committee for Electrotechnical Standardization (CENELEC), the European Telecommunications Standards Institute (ETSI) and selected EC directives, call the EC hotline at (301) 921-4164 (updated weekly).

The General Agreement on Trade and Tariff (GATT) hotline at (301) 975-4041 provides information on proposed regulations which affect foreign trade.

U.S. exporters may also want to contact the nearest local office of the U.S. and Foreign Commercial Service (US&FCS) located in 68 U.S. cities and 68 foreign countries.

Companies that have experienced a recent decline in sales and employment may be eligible for government assistance of up to 75 percent of the cost of consulting services relating to ISO certification. Interested parties should contact one of the twelve Trade Adjustment Assistance Centers (TAAC) listed in Table A.1

Table A.1 Trade Adjustment Assistance Centers

New England TAAC • 120 Boylston Street Boston, MA 02116 Tel: (617) 542-2395	Southeast TAAC • Georgia Institute of Technology Research Institute Economic Development Lab. Atlanta, GA 30332 Tel: (404) 894-6106	Mid-West TAAC • Applied Strategies Intl. 150 N. Wacker Drive Suite 2240 Chicago, IL 60606 Tel: (312) 368-4600
New Jersey TAAC • NJ Economic Dvlpmnt Authority Capital Place One-CN 990 200 South Warren Street Trenton, NJ 08625 Tel: (609) 292-0360	Southwest TAAC • 301 South Frio Street Suite 225 San Antonio, TX 78207-4414 Tel: (512) 220-1240	Rocky Mountain TAAC • 3380 Mitchell Lane, Suite 102 Boulder, CO 80301 Tel: (303) 443-8222
New York TAAC • 117 Hawley Street Binghamton, NY 13901 Tel: (607) 771-0875	Mid-America TAAC • University of Missouri at Colombia University Place, Suite 1700 Colombia, MO 65211 Tel: (314) 882-6162	Northwest TAAC • Bank of California Center 900 4th Avenue, Suite 2430 Seattle, WA 98164 Tel: (206) 622-2730
Mid-Atlantic TAAC • 486 Norristown Road Suite 130 Blue Bell, PA 19422 Tel: (215) 825-7819	Great Lakes TAAC • University of Michigan School of Business Administration 506 East Liberty Street Ann Arbor, MI 48104-2210 Tel: (313) 998-6213	Western TAAC • University of Southern California 3716 S. Hope Street, Suite 200 Los Angeles, CA 90007 Tel: (213) 743-8427

Source for Table A.1: Quality Systems Update Volume II, Number 7 (July 1992, p. 6).

Appendix B: Acronyms

ASQC: American Society for Quality Control (800) 248-1946

ANSI: American National Standards Institute (212) 642-4900

CEN: European Committee for Standardization

CENELEC: European Committee for Electrotechnical Standardization

DIS: Draft International Standard

EFTA: European Free Trade Association

EOTA: European Organization for Technical Approval

EOTC: European Organization for Testing and Certification.

The EOTC's role is to promote the mutual recognition—in the nonregulated product sector of the EC and EFTA—of conformity assessment activities which includes test results, certification procedures, and quality systems assessments.

EQS: European Committee on Quality Assessment and Certification. The EQS's role is to support mutual recognition of certificates.

ETA: European Technical Approval.

An interim organization which evaluates and submits technical directives to the EOTA.

ETSI: European Telecommunication Standards Institute

ISO: International Organization for Standardization (known as ISO)

NIST: National Institute of Standards and Technology (301) 975-4040

OIML: International Organization for Legal Metrology

Provides guidance on the applicability and use of the ISO 9000 Standard Series in the manufacture of measuring instruments

SIMIS: Single Internal Market Information Service (202) 377-5276

Appendix C: National and International Registrars (May 1992)

Registrar	Accreditation
ABS Quality Evaluations, Inc. 263 North Belt East Houston, TX 77060 (713) 873-9400	Registrar Accreditation Board (**RAB**) and Raad voor de Certificatie (**RvC**)
American Association for Laboratory Accreditation (A2LA) 656 Quince Orchard Road #304 Gaithersburg, MD 20878-1409 (301) 670-1377	
American European Services, Inc. (AES) 1055431 31st Street, NW, Suite 120 Washington, DC 20007 (202) 337-3214	Member of SEQUAL a subsidiary of APAVE which consults on certification with the French Association Française Assurance Qualité (**AFAQ**)
American Gas Association Laboratories 8501 E. Pleasant Valley Rd. Cleveland, OH 44131 (216) 524-4990	Memorandums of Understanding (MOU) with a number of European Accrediting Bodies.
Asociation Espanola de Normalizacion y Certification (AENOR) 28010 Madrid, Spain 410-4851	Certifies companies in Spain
AT&T Quality Registrar 1259 S. Cedarcrest Blvd Allentown, PA 18103 (215) 770-3285	RAB
British Standards Institution (BSI) Quality Assurance P.O. Box 375 Milton Keynes MK14 6LE United Kingdom 0908-220-908	National Accreditation Council for Certification Bodies (**NACCB**) and RVC
Bureau Veritas Quality International (BVQI) 509 North Main Street Jamestown, NY 14701 (716) 484-9002	RvC and NACCB

Canadian General Standards Board (Office) 9C1 Phase 3 Place du Portage, 11 Laurier St. Hull, Quebec (819) 956-0439	MOUs with European Accrediting Bodies
Det Norske Veritas Industry 16340 Park Ten Place, Suite 100 Houston, TX 77984 (713) 579-9003	RvC, NACCB, UNICEI (Italy) and SWEDAC (Sweden)
French Quality Assurance Association (AFAQ) Tour Septentrion Cedex 9 92081 Paris-La Defense France (33) (1) 47-73-49-49	Certifies companies in France
Intertek 9900 Main Street, Suite 500 Fairfax, VA 22031 (703) 476-9000	RvC
Lloyd's Register Quality Assurance (LRQA) 33-41 Newark Street Hoboken, NJ 07030 (201) 963-1111	NACCB and RvC
National Standards Authority of Ireland (NSAI) Certification Division Eolas, Glasnevin Dublin 9, Ireland 37-0101	Operates under the Irish Science and Technology Agency
Quality Management Institute Suite 800 Mississauga Executive Center Two Robert Speck Parkway Mississauga, Ontario, Canada L4Z 1H8 (416) 272-3920	MOU with European bodies
Quality Systems Registrars, Inc. 1555 Naperville/Wheaton Rd. Naperville, IL 60563 (708) 778-0120	RAB and RvC
SGS Yarsley Quality Assurance Firms 1415 Park Avenue Hoboken, NJ 07030 (201) 792-2400	NACCB

TUV Rheinland of North America, Inc. (TUV) 12 Commerce Road Newtown, CT 06470 (203) 426-0888	Germany's DAR and RvC
Underwriters Laboratories, Inc. (UL) 1285 Walt Whitman Road Melville, NY 11747-3081 (516) 271-6200	MOU with British Standards Institution (BSI)
Vincotte USA, Inc. 10497 Town & Country Way, Suite 900 Houston, TX 77024 (713) 465-2850	

Appendix D: Quality Manual

The following sample quality manual is an amalgamation of several quality manuals. The primary influence comes from a hardware-software company, but other industries are represented. As always, ideas presented in this manual are merely suggestions. In that respect this quality manual is nothing more than a model and should **not** be copied verbatim as is often done with manuals bought from software industries.

I have inserted comments in *italicized letters* followed by some samples. Most, but not all, sections have some sample text. The best quality manual is the one that truly reflects your company's policy and philosophy. The format suggested in this manual is one of many formats available to the user. There are basically three types of formats for quality manuals:

- The format adopted here. This format parallels the ISO standard by keeping the same structure and paragraph titles.

- The generic format. This format, often found in quality manuals overseas (mostly the U.K.) and adopted in the U.S., also parallels the ISO structure but the contents are much more generic. In fact, the contents are so generic that some enterprising individuals have included various software versions and are selling the diskettes for anywhere between $25 and $100. Three sample pages are included below.

- The third format is usually developed by the company that takes great pride in announcing that they have developed the manual without the assistance of a consultant. Of course, there is nothing wrong with that approach except to say that the quality manuals that I have seen tend to be much too verbose and thus too long and are often written in the future tense ("Document shall be controlled," . . . . "Test procedures will be implemented." . . . etc.). These manuals sometimes mistakenly include tier two and tier three information .

Since there is no current quality manual standard—and hopefully there will never be—you are of course free to use whichever format you prefer; however, I would strongly encourage you not to include tier two and three documents in your tier one manual. To do so will unnecessarily complicate your life and that of the registrar. Please note that the following pages, including the sample quality manual listed below, do not show how the pages are controlled. You will have to insert on each page some header or footer of your choice indicating how the document or pages are controlled. This can be achieved in a variety of ways. You could, for example, include in a footer or header (shown below) the following information: the name of who approved the document or page(s), the pagination, the version (which could be the day/month/year the document is issued) and other additional information which you think might be relevant.

Approved by: *A variety of formats are available. The manual could be approved by CEO, and each section could be approved by VPs or Directors.*	**Version:** Date or some other scheme of your choice. Engineering document control usually come up with some interesting numbering schemes.
Issued By: *Department.*	**Page:**

For more specific examples see my book entitled *ISO 9000: Preparing for Certification.*

Finally, you might want to indicate the scope of your quality manual. For example, I am often told that, although the XYZ company currently makes only three product lines, a couple of "ancient" units are still manufactured. Naturally, for these "ancient" products, the documentation does not conform to the ISO 9000 series. Therefore, I am often asked, "Must we go back and upgrade all of our old documents?" Of course not, but be sure to specify that the quality assurance system you are developing/implementing is to apply to such and such product(s).

But what about products that are "caught" in the ISO 9000 implementation euphoria, half-way through their design cycles? How will you convince product managers and designers that they must now do things differently? Difficult indeed, particularly if you do not have the support of key people. Hopefully, your current practices are not too far removed from what the ISO 9000 standards require.

Sample Pages From a Generic Quality Manual

Corrective Action

1 Scope

This policy defines the requirements for corrective action review and implementation within.

2 Policy

All nonconformances shall be investigated. The investigation shall evaluate the cause and, where necessary, assign corrective and preventive actions to correct the nonconformance and prevent recurrence.

3 Responsibilities

3.1 All managers are responsible, to the extent dictated by their terms of reference, for ensuring that corrective actions are identified and implemented in a timely and effective manner.

3.2 The Quality Assurance Manager, together with the Manager concerned, shall be responsible for monitoring the effectiveness of such actions.

3.3 All Managers shall be responsible for ensuring that staff under their control are familiar with procedures which support the requirements of this policy, and for the evaluation and monitoring to ensure continued compliance with these documented procedures.

Design Control

1 Scope

This policy defines the system utilized by the XXX company to control and verify design so as to ensure compliance with specified requirements.

2 Policy

2.1 XXX shall research, develop, and undertake design activity within the predetermined framework, defined in the applicable quality procedures.

2.2 Research, development and design activities shall be planned and coordinated by suitably qualified personnel. Planning shall include, but not be limited to, the identification of interfaces, design simulation, test/verification procedures and staged design reviews. All ambiguities and conflicting requirements shall be resolved during the planning stage.

2.3 Design outputs shall be verified and documented in order to demonstrate conformance against specified requirements. These shall include regulatory and safety requirements.

2.4 Appropriate testing strategies shall be developed in order to demonstrate compliance against design specifications.

2.5 Changes and modifications to designs shall be reviewed and approved in accordance with documented procedures.

3 Responsibilities

3.1 Senior staff, identified with local procedures are responsible for establishing and maintaining the controls applicable to research, development and design activities.

3.2 Senior staff shall ensure interfaces are maintained between relevant departments and disciplines, in order that information and specifications are effectively communicated in order to meet design requirements.

3.3 The Quality Assurance Manager shall ensure, through regular reviews, that the research, development and design management systems are effectively implemented, and that the controls meet design requirements.

Process Control

1 Scope

This policy defines the system at the XXX company to control all processes incumbent in the provision of services or the manufacturer of assemblies and sub-assemblies. The system ensures the interchangeability, repeatability and reliability of the processes during manufacture and field utilization.

2 Policy

2.1 XXX shall develop and maintain a range of documentation and standards to ensure development, production processes, end-item use, service provision, repair and maintenance programs are carried out in a controlled and consistent manner.

2.2 All such documentation shall be generated on the assumption that the absence of such documentation would seriously affect quality and/or field utilization.

2.3 Requirements for the subsequent approval of each process shall be clearly defined within the Standard, Procedure or Manual.

2.4 Where processes cannot be fully verified by subsequent inspection and testing, compliance shall be established by in-process monitoring in accordance with specified requirements.

2.5 All Standards, Procedures and Manuals shall be subjected to formal change control.

3 Responsibilities

3.1 It is the responsibility of all relevant managers to identify any process that may warrant special controls.

3.2 Senior management shall be responsible for ensuring development and utilization of the Standards, Procedures and Manuals necessary to maintain process control at acceptable levels.

3.3 The Quality Assurance Manager is responsible for evaluating and monitoring the effectiveness of process controls.

Notice that the prose often rephrases the ISO standards. The **Scope, Policy** and **Responsibility** headings mimic the ISO standard (nothing wrong with that, it is in fact a good idea). Notice also that the above samples could essentially be adopted, almost verbatim, by any company within the same industrial sector. The above sample pages are substantially different in style from the sample quality manual included in this Appendix. There is less information in the generic sample pages, tasks will be done but the reader often does not know how. The general intent is basically to assure the readers that all of the requirements specified in the Standard (9001) will be addressed. The writer deliberately avoids specificity. The manual listed below is more specific. Both styles are quite acceptable.

The use of "shall" can sometimes have an unfortunate and unintended effect of implied futurity. One could wonder when the company in question "shall" implement the said policy. Of course, the use of "shall" can also imply determination or promise.

THE
KNOWITALL
CORPORATION

QUALITY MANUAL

Controlled Document Printed on Stamped Company Paper

Date:	Assigned to:	Approved by:
Version:	Department:	Page 1 of 25

Table of Contents

Page

Revision History
0.1 Scope
0.2 Distribution and Manual Control
0.3 Glossary
0.4 Company History
0.5 KNOWITALL Mission Statement
0.6 Organization Chart
1.0 Management Responsibility
2.0 Quality System
3.0 Contract Review
4.0 Design Control
5.0 Document Control
6.0 Purchasing
7.0 Purchaser Supplied Product
8.0 Product Identification and Traceability
9.0 Process Control
10.0 Inspection and Testing
11.0 Inspection, Measuring, and Test Equipment
12.0 Inspection and Test Status
13.0 Control of Nonconforming Product
14.0 Corrective Action
15.0 Handling, Storage, Packaging, and Delivery
16.0 Quality Records
17.0 Internal Quality Audits
18.0 Training
19.0 Servicing
20.0 Statistical Techniques

Table 1. Revision History

Rev #	Date	Sections Changed	Summary of changes	Authorized by:
0.1	8/15/92	All	First Draft/Sample	J.L.L

0.1 Scope

Sample

This quality manual describes the quality assurance system of the KNOWITALL Corporation as required by the ANSI/ASQC Q9001/ISO 9001 standard. It solely covers all products manufactured at the Issaquah, Washington, U.S.A location and does not include products made by other U.S. sites or wholly owned overseas subsidiaries.

0.2 Distribution and Manual Control

Explain how and/or where controlled copies of the manual are distributed and maintained (i.e., how are obsolete copies removed?). It is also a good idea to state that photocopies of the manual are not controlled. Do not include names but rather titles, this makes it easier to control the manual as individuals are promoted or transferred.

Sample

Controlled copies of this manual are printed on special company paper and are provided for every manager, director, V.P and the C.E.O—see distribution list (Table 2.). Uncontrolled copies of the manual may also be sent to customers.

Table 2. Distribution List

Title	Department/Copy #	Location
C.E.O	Headquarters/# 1	Office #:
Quality Director	Quality Dept./# 2	Office #:
Chief Engineer	Engineering/# 3	Office #:
etc.		

0.3 Glossary

Many quality manuals include acronyms that may not always be familiar to the reader. It is therefore a good idea to define all acronyms (see sample).

Sample

Term	Definition
DOA	Dead on Arrival. A product that is delivered to a customer and cannot be used. This includes missing pieces, wrong product order, component failure, or any other cause leading to product malfunction.
ECN	Engineering Change Notice. A process change that communicates changes to participating departments.
ECR	Engineering Change Request. A form filled out by anyone who wishes to initiate a product change.
MIS	Management Information System.
PDG	Product Development Guide.
QAT	Quality Action Team. Team responsible for addressing corrective actions and/or process improvements.
SOP	Standard Operating Procedure. Tier three document describing a process, a procedure or assembly/work instructions.

0.4 Company History

It is a good idea to include a brief statement not to exceed a page, describing the company's origin, geographic location and type of products and/or services provided.

0.5 KNOWITALL Mission Statement

Most companies also include their mission statement.

A Comment Regarding Mission Statements

Most quality manuals include, within the first few pages, a mission statement, which in some cases, is apparently meant to substitute for the company's quality policy. I have often found these "Mission Statements" to be rather meaningless. In fact, as Russell L. Ackhoff points out, "Most corporate mission statements are worthless," mostly because they consist of "pious platitudes."[1] Ackhoff astutely observes that a mission statement should not "commit a firm to what it *must* do in order to survive but to what it *chooses* to do in order to thrive."[2] If you must include a mission statement, you might want to consider Ackhoff's wise suggestions:

- *It (mission statement) should contain a formulation of the firms objectives that enables progress toward them to be measured.*
- *A company's mission statement should differentiate it from other companies.*
- *A mission statement should define the business that the company* wants *to be in, not necessarily* is *in.*
- *A mission statement should be relevant to all the firm's stakeholders.*
- *A mission statement should be exciting and inspiring.*[3]

---

[1] Russell L. Ackhoff, *Management in Small Doses* (John Wiley & Sons, New York: 1986), p. 38.
[2] Ackhoff, op. cit., p. 38.
[3] Ackhoff, op. cit., pp. 38-41.

0.6 Organization Chart

A high-level organization chart is often included to help the reader better understand the company's structure. More detailed organizational charts can be included within each department's tier two. For some companies who seem to experience reorganizations every nine to twelve months, the inclusion of a high-level organization chart may not be wise, since it would require updating (and thus controlling) the changes at least once a year.

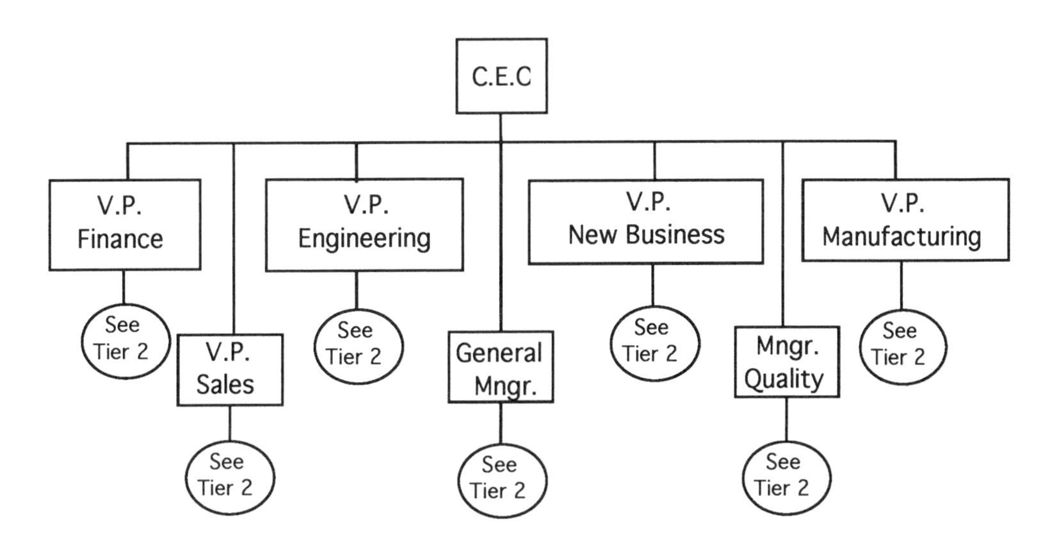

Sample

1. 0 Management Responsibility and Authority

1.1 Quality Policy

Include your company's quality policy.

Sample

The KNOWITALL company is committed to achieving customer satisfaction by ensuring that everyone understands and meets customer requirements. Our belief is that our success can only be measured by the continuing success of our customers. This policy is implemented on the production floor and is continuously monitored through the highest levels of management.

KNOWITALL's philosophy of continuous improvement relies on an intensive training program which emphasizes employees' direct involvement. Our aim is to respond to our customers' ever-changing needs in the most cost-effective way.

Signed:_________________________________, CEO

Communication of Quality Policy

All new employees attend a two-day workshop on continuous improvement. In addition, departmental workshops designed to emphasize the continuous improvement process are conducted at least once a year.

1.2.1 Responsibility and Authority

An efficient way to address this sub-paragraph is to include all quality responsibilities in a table format (see sample).

Sample

Responsibilities for implementing and maintaining the quality system are shown in Table 3.

Table 3 Quality Responsibilities

Title	Responsibility	Reports to
CEO	Leadership and staffing	Board of Director
CQI Manager	ISO 9001 compliance	VP Engineering
Director of Engineering	Product validation Product documentation Product identification and traceability	VP Engineering
Director of Tech. Services	Document control	VP Engineering
Sales Manager	Monitoring customer satisfaction	VP Sales
Customer Support Mng.	DOAs analysis Customer notification Factory contact	Director of Sales
Production Manager	Nonconforming products Work instructions	VP Manufacturing
Materials Director	Supplier quality	VP Manufacturing
Operations Manager	ISO 9001 compliance	VP Manufacturing
Dir. of Int'l Sales	ISO 9003 compliance	VP Sales
Doc Control Supervisor	Document control	Dir Tech Services
Application Engineers	Report problems via customer contacts	Customer Support Manager
Change Board	Manage product change	Doc Control
All Employees	Responsible for reporting nonconforming product via ECRs	Immediate Supervisor

1.2.2 Verification Resources and Personnel

Several options are available; you can either identify verification activities by product or by function (see sample).

Sample

Verification activities are performed throughout the product cycle. Resources are determined as part of the annual planning process. Trained personnel are recruited and hired per our standard hiring practice (see Human Resources tier two).

Design Process	Design reviews are done on all products. (For further details see Design Review tier two and tier three.)
Design Hand-off	A validation checklist is maintained by Engineering (see tier two).
Validation	The validation department tests products against customers briefs.
Production	Production tests follow Test Engineering specs (see Production tier two and three).
Eng. Change	ECNs follow a review process (see Design tier two).
Quality System	Internal audits are performed twice a year (see Internal Audit tier two).

References

1) Human Resources tier two and four (records)

1.2.3 Management Representative

State who is responsible for the maintenance of the ISO standard.

Sample

The Manager of Quality is responsible for ensuring that the ISO 9001 (2 or 3) is implemented and maintained. This responsibility includes:

- Ensuring that the requirements are implemented, maintained and understood.

- Coordinating quality related activities.

- Ensuring that internal audits are scheduled, and carried out.

- Ensuring that internal audit reports and corrective actions are reported to management (see 1.3 Management Review).

1.3 Management Review

Explain how often the quality system will be reviewed. Describe the process or refer to a tier two document which explains the process.

2.0 Quality System

Summarize the format of your quality system and quality plans.

Sample

The quality manual is the primary document of the quality system. In addition, each department maintains its own tier two and tier three documentation. Quality plans are maintained by the Quality Department. Implementation and maintenance of the quality system is the responsibility of all managers and line employees.

3.0 Contract Review

Describe the contract review process which addresses the three main points of the ISO standard. For multiple special products, this section can consist of several paragraphs. For firms offering standardized or "catalogued" products, or one or two products, this section may consist of only one or two paragraphs.

4.0 Design Control (9001 only)

Be sure to address all sub-clauses (General, Design and Development Planning, Activity Assignment, Organizational and Technical Interfaces, Design Input, Design Output, Design Verification and Design Changes). The following sample is adapted from a high-tech software-hardware company.

Sample

4.1 General

Design Engineering Managers and Project Leads are responsible for ensuring that specifications negotiated by the Marketing department are met. The Project Management department coordinates cross-departmental team meetings and project schedules. See relevant second tier documents (*more specific document references could be included*).

4.2 Design and Development Planning

Engineering management is responsible for ensuring that design and verification activities are planned and assigned to qualified staff with adequate resources. A master staffing plan and information from Product Specification are used to determined resources required.

4.4.2.1 Activity Assignment

The Project Leads keep plans for activity assignments. These plans are updated as necessary and monitored by the Program Manager.

4.4.2.2 Organizational and Technical Interfaces

The management of all related engineering and engineering support functions meets in regular quality improvement meetings designed to facilitate technical interfaces.

4.4.3 Design Input

Input requirements are provided by Product Marketing. Conflicting requirements are resolved by Design Engineering and Product Marketing.

4.4.4 Design Output

The validation department verifies that the product matches the manual. Various industry standards are used to identify and classify problems.

4.4.5 Design Verification

Verification is performed at several stages (see 4.1.2.2) .

4.4.6 Design Changes

Design changes approved by Design Engineering are communicated to various departments via electronic mail.

References

1) Design Engineering tiers two and three
2) Marketing tier two
3) Contracts

5.0 Document Control

This section requires that you specify how documents are controlled (i.e., who issues and approves what document, how they are updated, etc.)

Sample

Products are controlled via an ECN (Engineering Change Notice) system. Authorization requirements vary, and are documented either in a document, or in the procedure controlling it.

Individuals receiving controlled documents are responsible for insuring that they replace their old copy with new ones. Revisions are made on an as-needed basis. The following table lists all quality control documents, who issues them and how they are controlled (see Table 4).

Table 4 Controlled Documents

Quality Document	Issued By	Controlled By
Quality Manual	Quality Manager	Document Control
Raw Material Specs.	Director of Materials	Materials
SOPs	Department Managers	Document Control
Human Resources	Managers	Human Resources
Work Instructions	Production Manager	Production Manager
Engineering Change Notice	Document Control	Document Control
Internal Audit Procedures	Quality Manager	Document Control
Product Specifications	Project Lead, Marketing	Program Manager
Engineering Drawings	Engineering	Document Control

6.0 Purchasing

This is a rather straight-forward section which focuses on:

6.2 Assessment of Sub-contractors

Sample

Qualified suppliers of critical raw materials and services must demonstrate capability to meet all requirements as outlined in the purchase contract. Vendor ratings are maintained with the purchasing department (see purchasing tier four).

6.3 Purchasing Data

Sample

All purchasing documents are reviewed for accuracy (see purchasing tier four for samples).

6.4 Verification of Purchased Product

For some companies this is one of the few paragraphs that might not be applicable. The paragraph refers to your customers ability to verify at source product. This clause also includes toll processes.

References

1) Purchasing tiers two and three

7.0 Purchaser Supplied Product

Another clause which might not be applicable. The clause addresses itself to suppliers who incorporate components provided by their customers (i.e. the purchaser) into the final product.

8.0 Product Identification and Traceability

Contents will vary depending on contract requirements and industry. For example, the following sample would certainly not be acceptable for the pharmaceutical or aerospace industries.

Sample

There are no customer requirements for identification and traceability. A serial number is assigned to each unit. This number is tied to the customer records. Component changes are reflected in the higher-level part number.

References

1) Quality Records
2) Purchasing Order

9.0 Process Control

9.1 General

Sample

Each department manager is responsible for the development and implementation of the process control procedures in his department. Process approval, and control/monitoring of key parameters is identified in the department's Standard Operating Procedures.

Product teams meet regularly to review work instructions and criteria for workmanship. Equipment approval is monitored by maintenance (see maintenance tiers two and three).

9.2 Special Processes

Identify if any.

References

1) Departmental SOPs
2) Maintenance tiers two and three

10.0 Inspection and Testing

10.1 Receiving Inspection and Testing

10.1.1 *Some companies rely on their suppliers, others follow 100 percent inspection, yet others still rely on various acceptance sampling plans. The contents of this paragraph will affect the next paragraph (how special releases are handled and recorded).*

10.2 In-Process Inspection and Testing

Sample

In-process inspection and testing is performed at key manufacturing/assembly points along the process. Nonconforming product is identified to await proper disposition (see 13.0 and 14.0).

10.3 Final Inspection and Testing

Sample

Prior to shipment, final inspection and testing are performed as per quality plan requirements and laboratory procedures.

10.4 Inspection and Test Records

Sample

Records of all inspection and test records are maintained by the quality department and the laboratory.

References

1) Laboratory procedures
2) Quality plan
3) Shipping records

11.0 Inspection, Measuring, and Test Equipment

Be sure to carefully read this clause, it is the longest clause and will require careful implementation. The implementation of clause 4.11 should be commensurate with the type of product and regulation called for either by the customer or federal agencies. It is an important clause because one cannot control processes without controlling instruments. Be sure to distinguish between laboratory instruments (often used to verify specifications) and process instruments used to ensure that the final product is within specification. You do not need to monitor (calibrate, maintain, etc.) all process gauges and instruments (this is likely to be prohibitively expensive anyway). However, you will have to monitor some instruments. For some companies, "some instruments" might mean a couple of dozen gauges, scales or torque screws, for others it might mean a couple of thousand gauges! (see Chapter 3).

12.0 Inspection and Test Status

How do you identify/record conforming or nonconforming products throughout production?

13.0 Control of Nonconforming Product

Sample

Nonconforming products are segregated for review. There are basically three types of causes leading to nonconforming products: manufacturing errors, test procedures and design errors.

13.1 Nonconformity Review and Disposition

Sample

For each of the above causes the appropriate department is advised of the problem(s) and is responsible for proper disposition and/or solution. Design errors are monitored/recorded via ECNs. Manufacturing error are monitored/recorded via DMRs (Discrepant Material Report). All nonconforming products that are not scrapped are re-inspected.

References

1) ECNs
2) DMRs

14.0 Corrective Actions

Sample

Evidence of unsatisfactory quality of product or service may be obtained from customer complaints or from information derived from internal tests or inspections. Corrective actions are specified in Table 5.

Table 5 Corrective Actions

TYPE OF PROBLEM	DISCOVERED BY	CORRECTIVE ACTION
DOA Report	Field Office Applications	See Tier Three See Tier Three
Incident Report	Validation	See Design Engineering Tier Three.
Manufacturing	Product Team	SOPs

References

1) Design Engineering tier three
2) Field Office reports
3) Manufacturing tier three

Note: Some registrars have a strict interpretation of subparagraph b) which reads as follows:

"b) analyzing all processes, work operations, concessions, quality records, service reports, and customer complaints to detect and eliminate *potential causes* of nonconforming product;" (ANSI/ASQC Q91 and Q92 paragraph on *Corrective Actions*, emphasis added).

I have been told by a client that a registrar wanted to know what procedure was in place to address subparagraph b). I do not really know how one can detect and eliminate potential causes of nonconformity prior to their occurrence, unless of course, one possesses a crystal ball! Naturally, one could identify potential causes but, after all, identification does not guarantee occurrence and thus may not require any action. Such analyses, should have been anticipated, or at least studied, during the design phase. Still, although no doubt well intended, the choice of the word "potential" is, I believe, unfortunate. Since you have been warned, you might as well address the issue. Also, remember that the 1994 version (on Corrective Actions) is even more demanding since *preventive actions* have now been added (see page 169).

15.0 Handling, Storage, Packaging and Delivery

The purpose of this section is to ensure that the product is properly handled (if applicable), stored (if applicable), packaged (if applicable) and delivered.

16.0 Quality Records

This paragraph can be perceived as the Standard's hub. One of the easiest ways to address it is to use a table format.

Sample

Table 6 Quality Records

Records	Purpose	Owner	Location and Retention
Product Specification	Define Requirements	Program Manager	Product File: 5 Years
Validation Incidents	Identify Problem	Validation	On-line Database: 5 Years
ECNs	Document Product Changes	Doc Control	Doc Control: 10 years
Etc.			

17.0 Internal Quality Audits

Briefly state who will conduct internal audits, how often and/who will be informed of audits. You will likely have to refer to a tier two audit procedure document; see Chapter 7 for further details.

18.0 Training

Sample

18.1 All new employees are required to attend an orientation session which emphasizes safety, quality, specific skills required to perform the job and State as well as Federal regulations.

18.2 Each level of management is responsible for the training of all subordinates in all aspects of the job they perform. Personnel performing specific assigned tasks are qualified on the basis of education, training or experience as required.

18.3 Formal training records are maintained in an electronic data system by the Human Resource Department. In addition, each department maintains its own department specific training file.

References

1) Human Resources

19.0 Servicing

This clause may not be applicable for some companies. If not, simply say so. If it does apply, elaborate. For some companies, servicing is such a major activity that it may require ISO 9002 or 9003 certification for the sub-divisions responsible for servicing.

20.0 Statistical Techniques

This last paragraph (one of the few "Where appropriate" clauses), may or may not apply to you. If your customers require you to use SPC or other related techniques (DOE for example), you will likely elaborate on the use of SPC on your processes. If you are in the business of one of a kind (or nearly so), you may not be able to refer to this paragraph. Be sure to explain why. You may want to investigate whether or not the use of statistical techniques can help you better monitor your processes. When used correctly and intelligently, statistical techniques can be very helpful.

Index

An index is always very valuable, particularly for auditors.

Appendix E: Suggested Answers to Case Studies

Suggested answers for Case Study 1

It is rather surprising how many ISO paragraphs apply to this brief scenario. There are at least eight paragraphs that could apply, they are (ISO 9001 references):

- 4.5 Document Control
- 4.9 Process Control (particularly (b))
- 4.10.1, 4.10.2 and 4.10.3 Inspection and Testing (Receiving, In-process and Final Inspection)
- 4.11 Inspection, Measuring and Test Equipment
- 4.12 Inspection and Test Status
- 4.13 Control of Nonconforming Product
- 4.14 Corrective Action
- 4.18 Training

Document control: There are plenty of documents available around the console. Since the Honeywell computer controls the process and hence the quality of the product, one would like to know how/if the Honeywell documents are controlled.

Process control: That should have been easy. The control room must be involved with process control. You might think that documented work instructions are required, however the screens provide the user with all the documentation that is required (assuming of course the user has been trained, see Training). One would like to know how the operator knows what to monitor, when and what action to take. Since the process is affected by sudden changes in the weather, how are these climatic accounted for to prevent degradation in product quality? In some closed-loop systems the computer automatically adjusts flow rates, temperature etc. Still, some human intervention is usually required. Where is this documented? Are the screens the documents? Probably, but how are the screens monitored (back to training!).

Inspection and Testing: We are told that raw materials are delivered via trucks and pipeline. How is it inspected? By whom? (Most likely the lab.) How frequently? How do you inspect pipeline product? What do you do if you don't like the product? In most cases you have little choice, you must accept it and do the best you can by adjusting your process. Could we see lab procedures and records of inspection?

The same approach would apply to in-process inspection. How often? What about the on-line analyzers? (See next paragraph.) How and where do you hold nonconforming product? The process industry is in some respect "fortunate" in that it can, in most cases, temporarily store nonconforming product for reprocessing. Of course that is expensive, but it's better than scrapping the product. In this particular case, you must remember that the splitter is only making one product. The unit is very efficient and everyone (including the software) really know what they are doing.

Final inspection questions should be asked at the lab. One would like to know what happens when the on-line analyzer does not match laboratory analysis.

Inspection, Measuring and Test Equipment: We are told that several instruments are monitored by the Honeywell. But do we know anything about these instruments. What about the maintenance crew? Do they maintain equipment or other units/equipment? How often? Are records kept? Can we see them? What is the frequency of inspection?

Don't get confused with laboratory instruments which must be under a strict calibration program. It is easy to ask all sorts of question about the precision of an instrument, but sometimes it is very difficult to access an instrument and therefore very expensive. Still, how do the operators know when an instrument has gone bad? (Usually, the answer is experience and common sense). If there is a steep temperature gradient between two adjacent points in a process, it does not take too long to figure out if it is the instrument or a broken pipe.

Since the computer sofware is used to monitor the process and thus the quality, one could ask: Who monitors the software and hardware? How was it tested (see last paragraph of 4.11)? All of this is usually well described in the supporting computer documentation.

Inspection and Test Status: Of course you can't expect process engineers to tag molecules, still they need to know what is good and where it is stored. Continuous processes are monitored continuously and usually shipped real time or stored in tanks prior to shipment. Tank storage are also inspected prior to delivery.

Nonconforming and Corrective Action: See above for answer. Corrective action, in the sense described in 4.14 (i.e., to prevent recurrence) is rarely done. Nonetheless, all facilities do implement corrective actions, but there is usually little or no record keeping or analysis for patterns of nonconformances.

Training: Obviously, when a process is controlled by a sophisticated system, one would like to know what sort of (recorded) training the operators have had. In this particular case, there was no problem.

Case Study 2

Several paragraphs would apply to scenario 2. For the first half of the scenario, one could refer to paragraphs 4.5 (ISO 9001, Document Control) and 4.9 (Process Control) as well as 4.10.2 (In-process Inspection). One could also have asked questions relating to paragraphs 4.13 and 4.14. Much of the same applies to the N/C department (4.5 and 4.9). One would also like to know about training (4.18). And what about maintenance? (See 4.9 c, which specifies "the approval of process and equipment, as appropriate"). As far as the changing of the drill is concerned, one could *suggest* the use of SPC to anticipate tool/drill change.

It seems that procedures are not followed or are improvised/modified without (perhaps) being updated/controlled on the master instructions. It is not surprising to learn that in this particular case, the company in question relies extensively on 100 percent in-house inspection (which is well known not to be 100 percent effective). Moreover, master documents are often updated by simply using white-out or glueing pieces of papers on "old" procedures. Signatures cannot be found. Of course, the reader should also know that the work force had been reduced by fifteen percent and production had *slightly increased.* Obviously, this "lean" system has to "crack" somewhere, somehow, sometime. Naturally, the implementation of an ISO 9000 type quality assurance system was not always accepted. Everyone knew (particularly manufacturing), that the already stressed system was about to be stressed further. Consequently, not every one wanted to take part in the ISO "program." This is often a dilemna that sooner or later must be faced by management. What are your priorities, how much do you really want to achieve registration, at what cost, and why?

Appendix F: ISO/IEC Guides

ISO/IEC Guide 2:1991. *General Terms and Their Definitions Concerning Standardization and Related Activities.*

ISO/IEC Guide 7:1982. *Requirements for Standards Suitable for Product Certification.*

ISO/IEC Guide 16:1978. *Code of Principles on Third-Party Certification Systems and Related Standards.*

ISO/IEC Guide 22: 1982. *Information on Manufacturer's Declaration of Conformity with Standards or Other Technical Specification.*

ISO/IEC Guide 23: 1982. *Method of Indicating Conformity with Standards for Third-Party Certification Systems.*

ISO/IEC Guide 25: 1990. *General Requirements for the Competence of Calibration and Testing Laboratories.*

ISO/IEC Guide 27: 1983. *Guidelines for Corrective Action to be Taken by a Certification Body in the Event of Misuse of the Mark of Conformity.*

ISO/IEC Guide 28: 1982. *General Rules for a Model Third-Party Certification System for Products.*

ISO/IEC Guide 38: 1983. *General Requirements for the Acceptance of Testing Laboratories.*

ISO/IEC Guide 39: 1988. *General Requirements for the Acceptance of Inspection Bodies.*

ISO/IEC Guide 40: 1983. *General Requirements for the Acceptance of Certification Bodies.*

ISO/IEC Guide 42: 1984. *Guidelines for a Step-by-Step Approach to an International Certification System.*

ISO/IEC Guide 43: 1984. *Development and Operation of Laboratory Proficiency Testing.*

ISO/IEC Guide 44: 1985. *General Rules for ISO or OEC International Third-Party Certification Schemes for Products.*

ISO/IEC Guide 48: 1986. *Guidelines for Third-Party Assessment and Registration of Supplier's Quality System.*

ISO/IEC Guide 53: 1988. *An Approach to the Utilization of Supplier's Quality System in Third-Party Product Certification.*

ISO/IEC Guide 54: 1988. *Testing Laboratory Acreditation Systems-- General Recommendations for Operations.*

ISO/IEC Guide 55: 1988. *Testing Laboratories Accreditation Systems. General Recommendations for Operation.*

ISO/IEC Guide 56: 1989. *An Approach to the Review by a Certification Body of Its Own Internal Quality System.*

ISO/IEC Guide 57: 1991. *Guidelines for the Presentation of Inspection Results.*

One should also note the Draft Document ISO/CASCO 170 (Guide 58), *Calibration and Testing Laboratory Accreditation Systems—General Requirements for Operation and Recognition (Revision of ISO/IEC Guides 38, 54 and 55).*

Appendix G: Summary of the "Working Document on Negotiations with Third Countries Concerning the Mutual Recognition of Conformity Assessment"[1]

The following is but a condensed version of some of the key paragraphs. The complete document is eight pages long.

1.3 "The Council Resolution of 21 December 1989 provides for the conclusion of mutual recognition agreements on conditions that:

- the competence of the third country bodies is and remains on a par with that required of their community counterparts;

- the arrangements are confined to reports, certificates and marks drawn up and issued directly by the bodies designated in the agreements;"

The resolution also states that ". . . the Parties have an equivalent guarantee of access to the market for the sector(s) covered by the agreement in terms of the requirements of the laws of the two Parties and the means of proof of conformity with these requirements."

1.4 ". . . all products placed on the Community market have a safety level equal to or greater than that laid down by these essential requirements."

1.5 "Community notified bodies will be able to certify the compliance of Community products with the requirements of the third country concerned. Conversely, third countries will be able to have the conformity of their products with Community requirements evaluated by notified bodies established on their territory."

---

[1] Commision of the European Communities, DOC. CERTF. 91/1 - Rev 3. Brussels, 18. 11. 91.

The document also refers to ". . . verification of the degree of conformity, in particular laboratory testing and quality checks carried out, with reference to EN 29000 and EN 45000 standards or *other standards shown to be equivalent,. . .*"

Paragraph 2.2 states that "The 'regulated' field concerns the products or sectors for which the public authority has decided to take action to defend the public interest on grounds of safety, health protection, environmental protection, consumer protection, etc., by requiring evaluation of conformity by third parties."

2.4 "The agreements will have to ensure that the technical competence and responsibility of the bodies instructed to carry out the conformity assessment in the third countries are and remain on a par with that required of their Community counterparts."

3.1.2.2 "An agreement between a Member State and a third country will have the effect that third country products which conforms to the provisions of the agreement are treated as products of the Member State concerned, as far as conformity assessment is concerned."

3.2.3 "The notified bodies of the third countries must take part in coordination meetings with the Commission and their Community counterparts."

Appendix H: Component Types Covered by IECQ

(Text extracted from "A Guide to The Worldwide Electronic Component Certification System. IECQ, Second Edition January, 1992)

The International Electrotechnical Commission Quality Assessment System (IECQ) is open to manufacturers of electronic components and independent test laboratories throughout the world (43 countries); it became operational in 1982. The IECQ secretariat is located with the Central Office of the IEC in Geneva, Switzerland.

A manufacturer in a participating country wishing to become approved under the IECQ system must first:

- Demonstrate to the satisfaction of the National Supervising Inspectorate (NSI, Underwriters Laboratories in the United States, for example), that his quality system meets the IECQ requirements as defined in the Rules of Procedure which include all applicable requirements of ISO 9001 or 9002.
- Indicate the generic, and sectional where relevant, specifications against which components will be manufactured.
- Nominate a Chief Inspector/Quality Manager.
- Prepare the necessary documentation as a basis for appraisal by the NSI.

Once Manufacturer Approval has been obtained the manufacturer has the option of proceeding to either Qualification Approval or Capability Approval.

Qualification Approval (QA) is granted to a manufacturer for an individual component or range of components. QA is granted on the acceptance of the test report which is representative of normal production and of the design and range of values.

Capability Approval (CA) obviates the need for the separate qualification of each component, as required by QA. CA is granted on the acceptance of test results of specified Capability Qualifying Components (CQCs) which must undergo the same processes and inspection procedures as are applicable to normal production.

List of components:

1. Attenuators, fiber optic
2. Branching devices, fiber
3. Cables, radio frequency
4. Capacitors, fixed
5. Connectors, low freq.
6. Connectors, rectangular
7. Ditto for, fibres + cables
8. Connectors, radio freq.
9. Inductors + transformers
10. Discrete semiconductors
11. Fibre optic, splices
12. Filters, ceramics
13. Integrated circuits
14. I.C. monolithic
15. Potentiometers
16. Printed boards
17. Quartz crystal unit
18. Relays, electromechanical
19. Resistors, fixed
20. Resistors networks
21. Resonators, ceramic
22. Semiconductors opto-electronic and liquid crystal devices.
23. Surge protective devices
24. Switches, electro-mechanical
25. Switches, keyboard
26. Thermistors
27. Tubes, electronic
28. Tubes generators
29. Varistors

Index

Aerospace Industries Association 194
Airbus' AIQI 32
American Association for Laboratory Accreditation 11
American National Standards Institute 3
American Petroleum Industry 10
Audit
  Case study 147
  IRS 163
  Repeatability and reproducibility 152
  Subjectivity 156
  Variability and standardization 157
Auditor
  Temperament types 155
Auditor's Psychological Types 154
Baldridge, Malcolm 24
Bellcore and ISO 9000-3 177
Boeing's D1-9000 32
British Computer Society 10
CE Mark 2, 15, 22
Change 71
  First and second-order 72
  Types of 72
Checklist 125
Contract review 41
Contract review and hotel 43
Contract review and software industry 43
Cross-Reference List of Quality System Elements 7
Cybernetics of ISO implementation 30
Deming 24
Department of Trade and Industry 10
Design and document control 46
Directives
  "L" and "C" series 11
  Food and beverages 207
  Medical devices 206
  Modules 20
  Telecommunications 205
  Telecommunications, medical and food and beverages 205
Document control 49
Documentation tiers
  Difference in scope 93
EC regulated products 2
EC's Technical Harmonization Directives 11
Electronic Components 18
EN 2000 10
EN 29000 2
EN 3042 10

EN 45000 14
EN 46001 10
Entropy, chaos and ISO 30
European Chemical Industry Council 10
European Committee for Electrotechnical Standardization 3
European Committee for Standardization 3
European Directives 14
European Norms 2
European Summit 1
Existing 9000 Series Standards 5
Facilitating ISO implementation 27
Federal Aviation Administration 10
Food and Drug Administration 10
Ford 50
Generic Process Flow Diagram 56
Good Manufacturing Practices 10
Houndshell 50
How Do You Interpret This Paragraph? 36
How to Implement Change 71
Implemetation strategy 33
Inspection, Measuring and Test Equipment 59
Internal Audit 121
  Closure 129
  How to ask questions 127
  How to conduct 123
  Human relations 128
  Preparation 123
  Reporting 128
  Shewhart cycle 122
  Team 123
International Electrotechnical Commission 3
International Organization for Standardization 3
Ishikawa 24
ISO
  10011 Series 5
ISO 45000 Series 6
ISO 9000
  Delegation 98
  Implementation 98
  Innovation 184
  Litigation 182
  Political economy 178
ISO 9000 and 21st Century 191
ISO 9000 and old documents 215
ISO Technical Committee 4
ISO/DIS 10012 65
ISO/IEC Guide 25 11, 65
ISO/IEC Guide 43 65
Kelvin, Lord 60

Leadership
Types of 79
Medical device 10
Mil Q 9858A 32
MIL-I-45208A 10
MIL-Q-9858A 10
MIL-STD-45662A 60
Mission Statement 228
Modules 17
NAMAS 11
National Accreditation Council for Certification Bodies 14
National Aeronautics and Space Administration 10
National Transportation Safety Board
Training 67
New and Related ISO Standards 5
Nonconformities 44
Notified bodies 14, 21
Null Hypothesis 153
Operating Procedures 115
OSHA
Training 68
Pannela, Robert 96
Pennella 65
Peters 24
Process control 52
Product certification 189
Product Liability Directive 1, 2
Product Sector Directives 16
Psychological Types
Auditors 154
Pyramid of Quality 92
Q-9000 Series 3
Quality records 66
service industry 66
Quality systems 24
Raad voor Certificatie 14
Regulated products 17
Risk
Producer and consumer 153
Second Order Change 73
Service industry 40
Singer 53
Single European Market 1
Single Internal Market, 11
Software Engineering Institute 49
Sydenham 61
TC 176 Committee 37,187
Technical Harmonization Directives 12
The "Show Me" mapping 32

Theories of change 78
Theory of change
  Argyris and Schon 80
  Hersey and Blanchard 79
Third-Party Audit 143
  Answering questions 144
  Formulating questions 146
TickIT 177
  Software 177
  U.K. 10
Tier three document 92
Tier two
  How to? 104
Tier two documentation 93
Total quality philosophy 23
Training 67
  Auditor 156
Training records
  Why? 68
Types of processes 39
Typology of ISO 9000 Applicants 37
What to document 54
Yourdon 49